Corporate Internal Reference
ELECTRONIC
CIRCUITS
Volume 1.1

Disclaimer

Corporate Internal Reference

ELECTRONIC CIRCUITS

Volume 1.1

LESERATI CIRCLE
A NORTH AMERICAN COMPANY
www.leserati.com

Corporate Internal Reference Electronic Circuits Volume 1.1

Published by the

LESERATI CIRCLE
A NORTH AMERICAN COMPANY
www.leserati.com

Copyright 2008 by
Leserati Circle, LLC
under a special agreement with
Intellin Organization, LLC
www.intellin.org

First year of publication 2008

Disclaimer:

The circuits, software or related documentation in this book are NOT designed nor intended for use (whether free or sold) as on-line control equipment in hazardous environments requiring fail-safe performance, such as, but not limited to, in the operation of nuclear facilities, aircraft navigation or communication systems, air traffic control, direct life support machines or weapons systems in which the failure of the hardware or software could lead directly to death, personal injury, or severe physical or environmental damage ("high risk activities")

The author(s) and publisher(s) take no responsibility for damages or injuries of any kind that may arise from the use or misuse of the circuits in this collection.

The author(s) and publisher(s) specifically disclaim any express or implied warranty or fitness for high risk activities. The circuits, software and related documentation are without warranty of any kind. The author(s) and publisher(s) expressly disclaim all other warranties, express or implied, including, but not limited to, the implied warranties of merchantability and fitness for a particular purpose. Under no circumstances shall the author(s) and publisher(s) be liable for any incidental, special or consequential damages that result from the use or inability to use the circuits and software or related documentation, even if he has been advised of the possibility of such damages.

Congratulations for having the second volume of ready-to-apply circuits. With this book, you got the luxury of being able to design and assemble electronic modules fast and worry free. It is a sure way to optimize satisfaction in your hobby. It will help you beat the competition if you are a professional electronic designer. Speed, efficiency, short development periods, error-free, user and maintenance friendly: these are the factors that are critical for success. This invaluable book filled with 101 practical ideas will help you beat project deadlines.

Make your ideas work!

Make your creativity pay. All that JUST IN TIME!

informative...
practical...
professional...
versatile...

Acknowledgments

Many Thanks to...

Engineer Mischa (Optical Recognition)
Engineer Salinger (Electronics)
Engineer P. Schmidt (Cybernetics)
Engineer N. Lay (Robotics)

INTRODUCTION

This collection is the second in a series. It contains 101 practical circuits grouped in ten general applications. Since most of the circuits are not limited to a single application, a circuit may have found its way into another group. This is one proof of the versatility of the circuits. Creativity needs versatility . You can combine several circuits into one large module to create a powerful electronic device especially designed for your exclusive project.

The table of contents lists the groups and titles of the circuits. The page number where a group begins can be found in this table. To find a particular circuit, turn to the group's beginning page. On this page, the circuits are again listed with their page numbers. To quickly find an application group, use the black markers on the edge of the pages. These markers coincide with the markers in the table of contents.

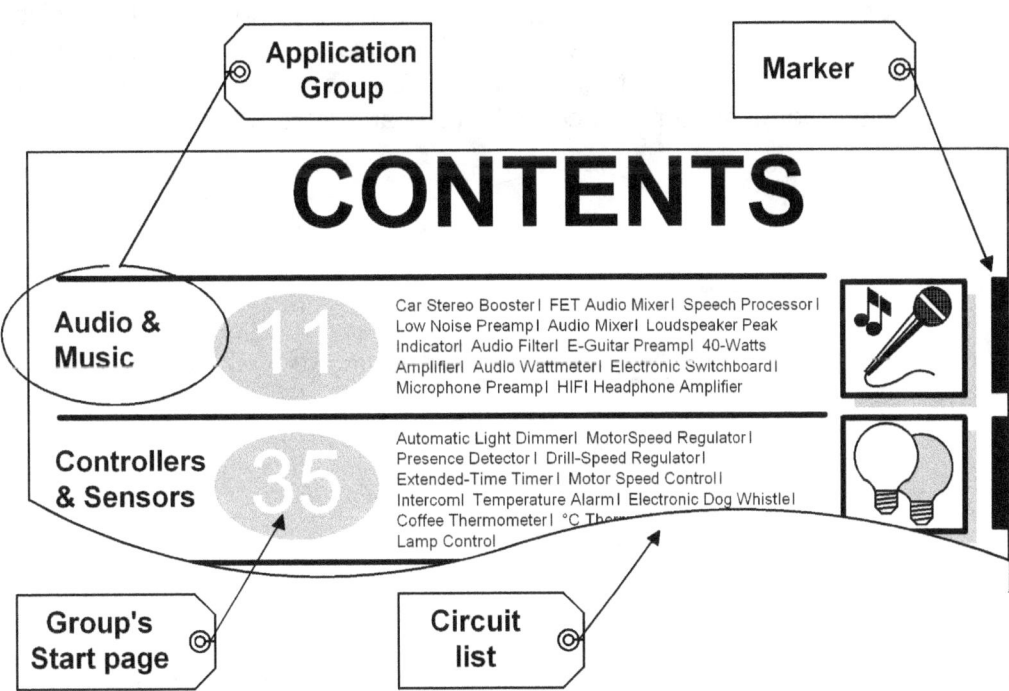

The transistors used in the circuits have more than one possible replacements. Their pin designations are also shown in details. This feature can help you avoid unnecessary delays. The pins shown are either in the bottom view or front view of the transistor unless otherwise noted. Large transistors which cannot or not planned to be installed directly on the PCB must be installed on a heatsink. A dashed circle around a transistor means the transistor must be heatsinked.

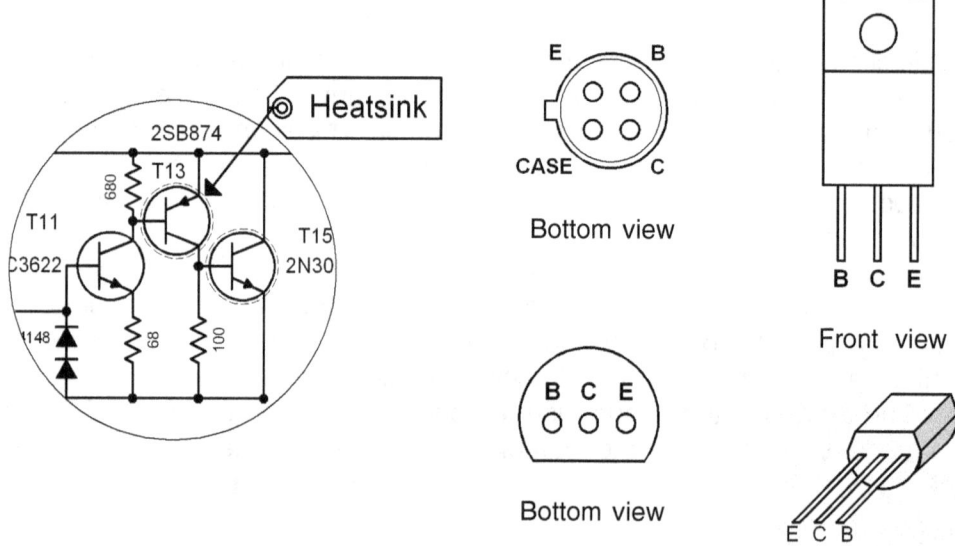

The capacitor values are given in microfarad unless otherwise specified. Electrolytic or polarized capacitors are marked with a plus sign in the diagram. This plus sign coincides with the capacitor's positive polarity in the circuit. Additionally, their voltage ratings are also given. Nonpolar capacitors are ceramic types and rated with 50 volts.

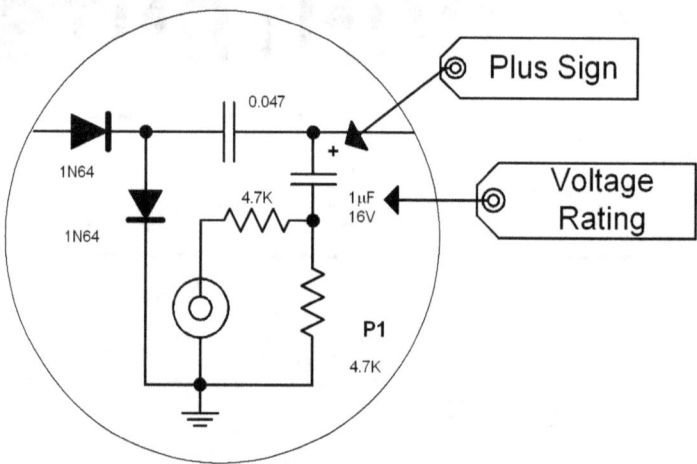

The resistor values are given in ohms (Ω), rated 1/4 watts, and are of carbon film type unless otherwise specified.

CONTENTS

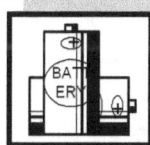

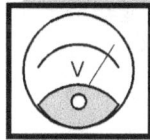

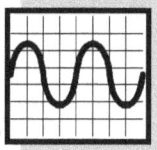

This page is intentionally blank.

AUDIO & MUSIC

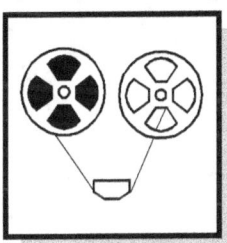

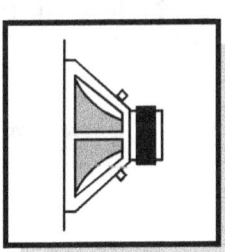

1   6.5 WATTS IC AMPLIFIER

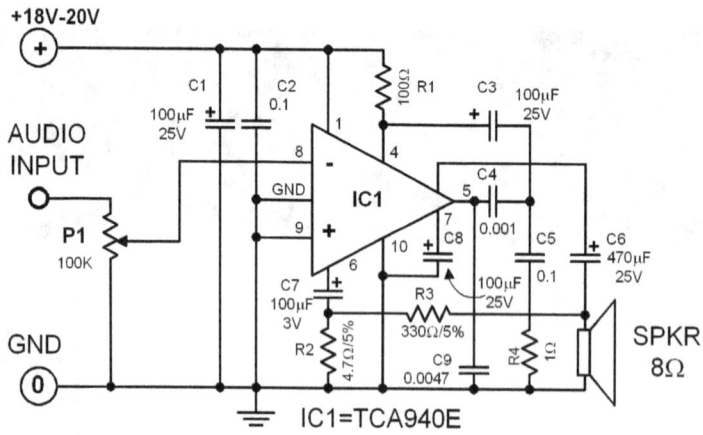

Diagram 1.0   6.5 Watts IC Amplifier

A medium powered amplifier for universal audio amplification purposes can be constructed with the monolithic amplifier TCA940E. This IC has negligible harmonics, low distortion, and a built-in short circuit protection. The IC's own heatsink must be soldered to the PCB's ground. Although its own heatsink is usually enough for most applications, it is a good idea to physically and thermally connect it to a large copper plate of the PCB. An area of around 4 cm^2 to 6 cm^2 is sufficient.

Technical Data

Power supply = 12 - 30 V
Power output = 6.5 W with 20V supply (8 ohm load)
                5.4 W with 18V supply (8 ohm load)
Frequency response= linear from 40 Hz - 20 kHz
Distortion factor (at 50mW to 3.5W) = 0.2%
Input sensitivity   =   110 mV
Input Impedance =   100K

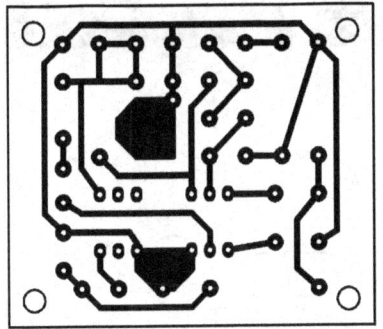

Figure 1.0 Printed Circuit Layout

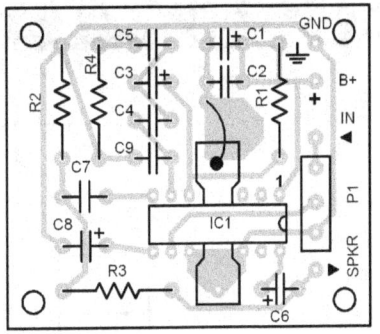

Figure 1.1 Parts Placement Layout

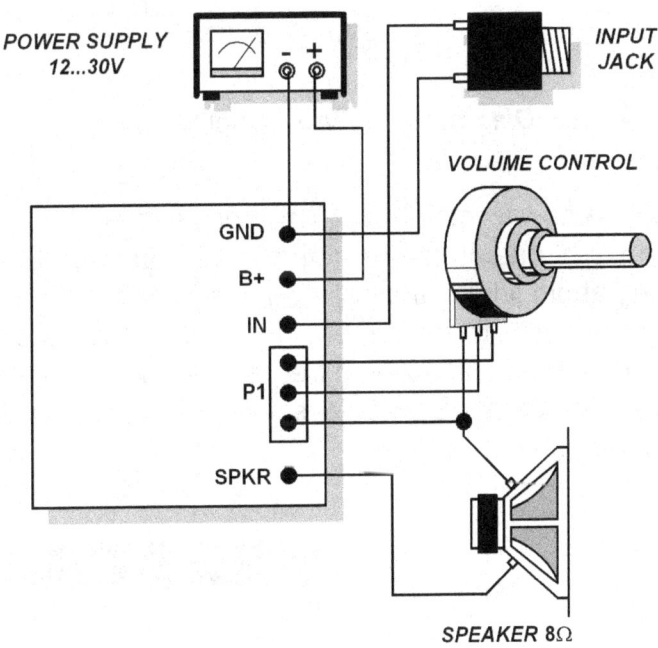

Figure 1.2 External Wirings

2 MINI AMPLIFIER

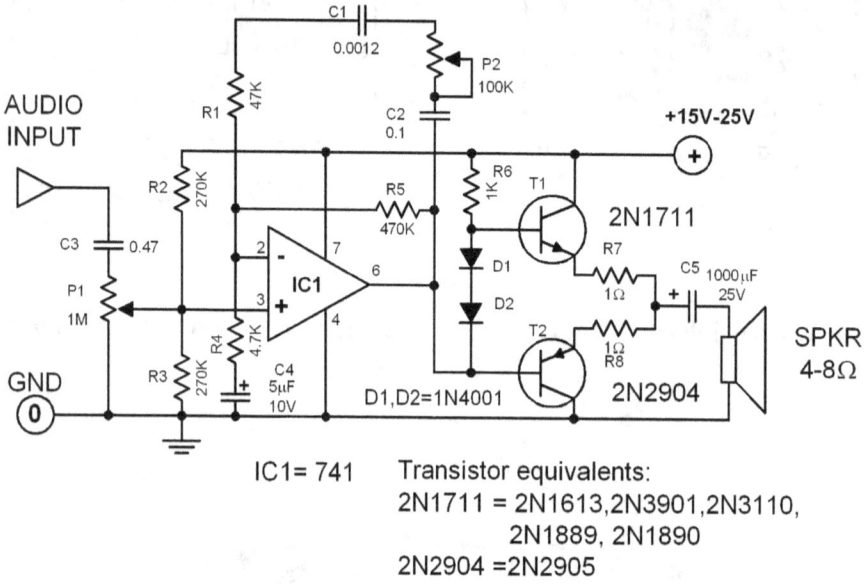

Diagram 2.0   Mini Amplifier

Small, simple, and easy to construct. This mini-circuit is a universal audio amplifier. Just the right one. It uses a standard 741 opamp IC which is readily available. No worry about finding the components. The output power can reach slightly more than 2 watts! Potentiometer P1 is the volume control while P2 controls the tone. P2 can suppress high frequencies up to maximum of 20dB. If you want to use transistors with 2S prefixes, select the transistor with specifications similar to the following:

	T1	T2
Type	npn	pnp
Collector Voltage (max)	75V	60V
Collector Current (max)	0.5A	0.6A
Power (max)	3 W	3 W
Current Gain	55	100-300
Transition Frequency(MHz)	>60	>200

Technical Data	
Input Sensitivity=	100mV
Power Output=	2W
Supply Voltage=	15V - 25V
Current consumption=	150 mA

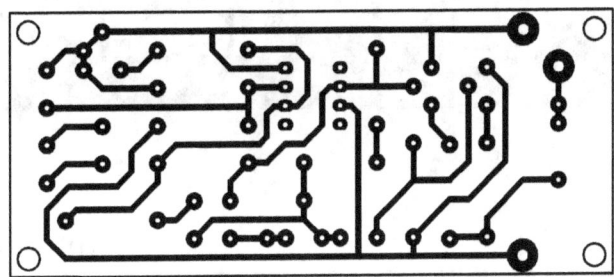

Figure 2.0 Printed Circuit Layout

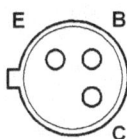

Bottom view

2N1711
2N3109
2N3110
2N1889
2N1890
2N2904
2N2905

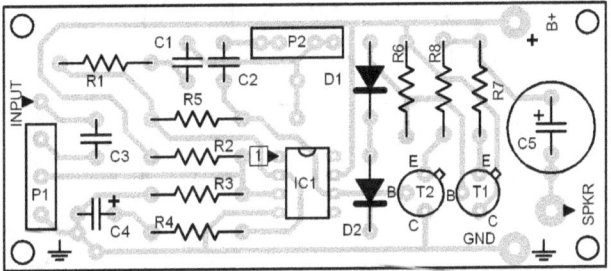

Figure 2.1 Parts Placement Layout

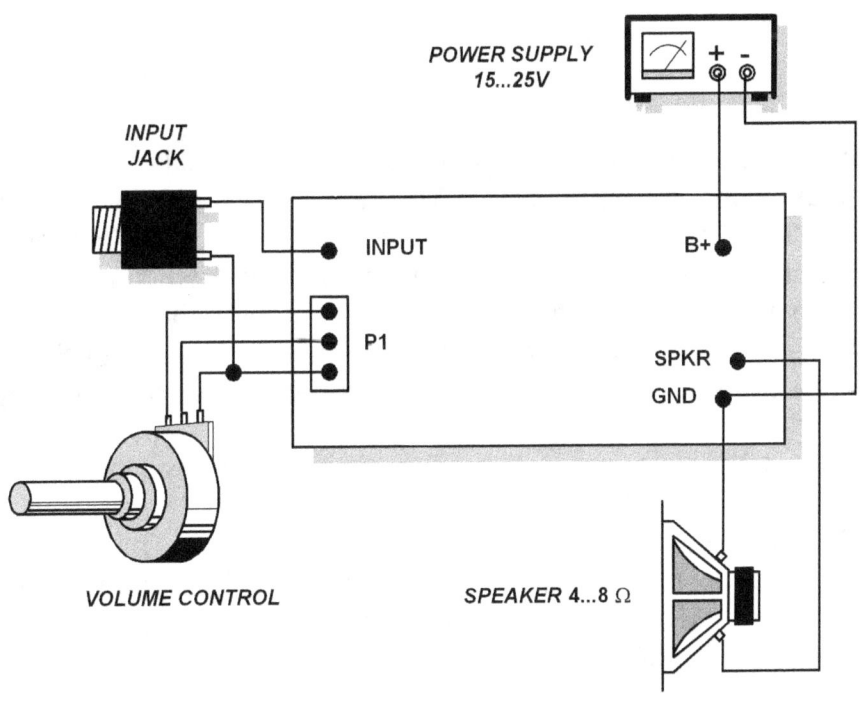

Figure 2.2 External Wirings

3 NOISE FILTER

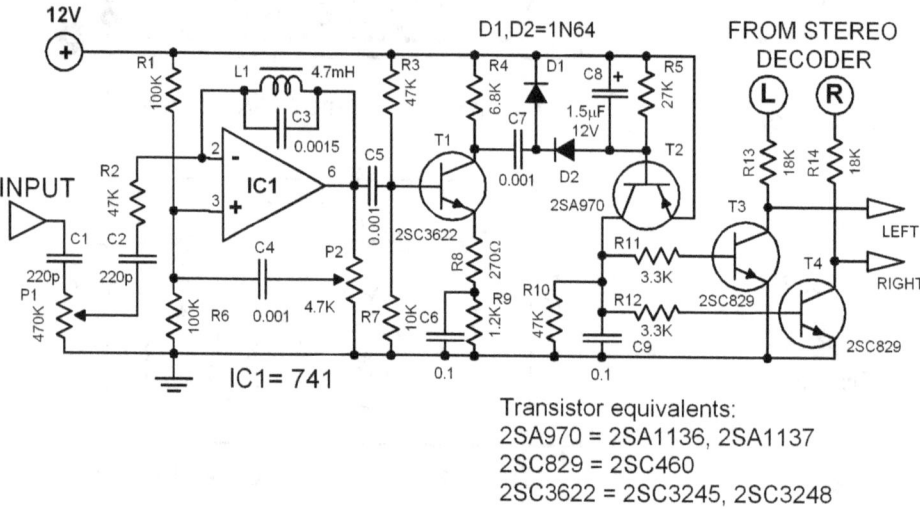

Diagram 3.0 Noise Filter

This circuit blocks the audio signal going to the final amplifier when the noise level exceeds a certain preset value. The noise threshold level can be set through potentiometer P1. This noise filter works with the "squelch" method - it prevents the amplification of noise signals by preventing them from reaching the amplifier. One practical application of the circuit is to mute the audio amplifier of an FM receiver when an empty frequency is found during channel changing or searching (the noise level is very high in an empty channel or frequency).

Installing the circuit is very easy. The noise filter is connected in parallel to the demodulator circuit of the FM receiver (see Diagram 3.2). The input point C1 is connected to the demodulator ouput of the stereo receiver. The potentiometer P2 sets the selectivity of the circuit. P1 sets the sensitivity level. Transistors T2 and T3 short circuit the output signal from the stereo decoder, preventing it from reaching the audio amplifier.

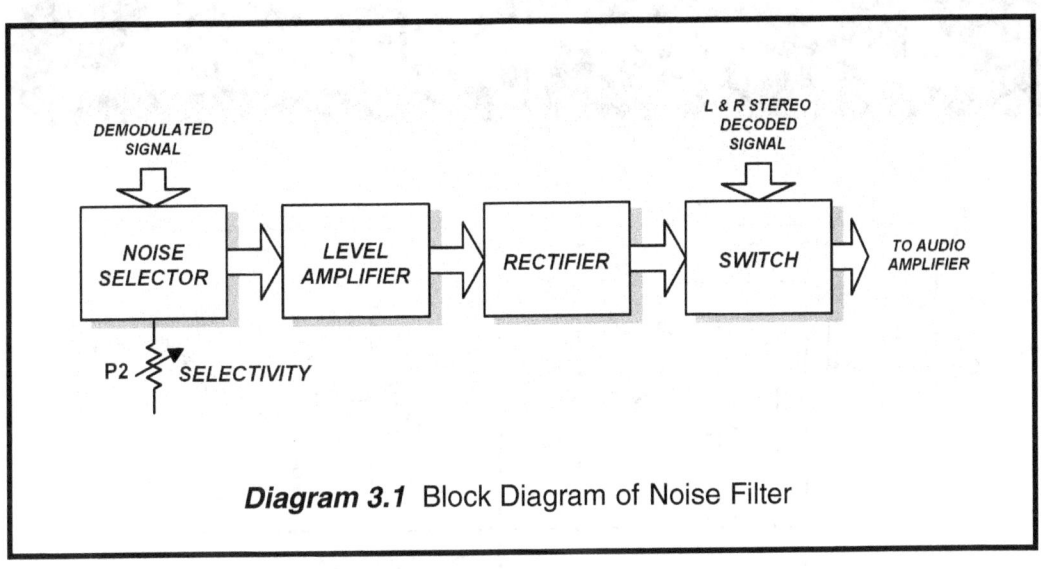

Diagram 3.1 Block Diagram of Noise Filter

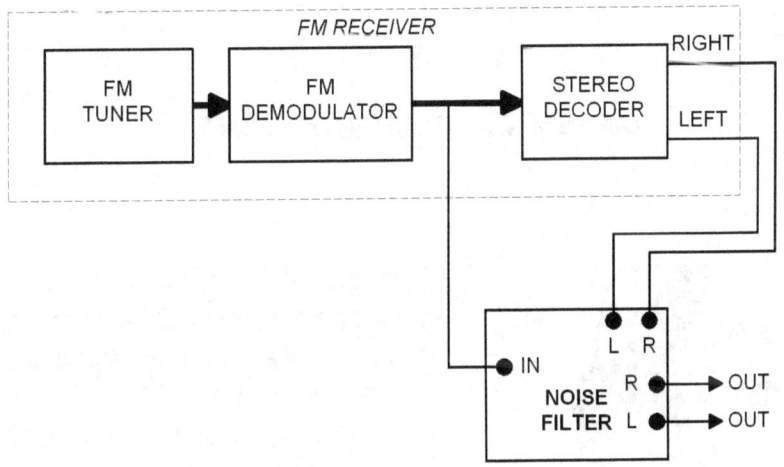

Diagram 3.2 Installation of Noise Filter

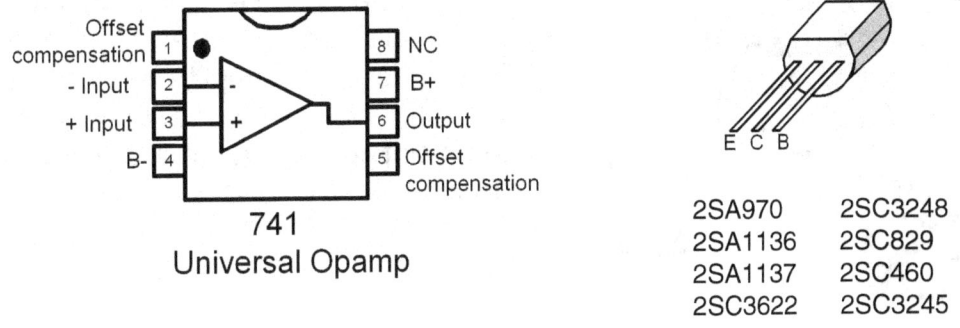

741
Universal Opamp

2SA970	2SC3248
2SA1136	2SC829
2SA1137	2SC460
2SC3622	2SC3245

4 GUITAR SOUND EFFECT

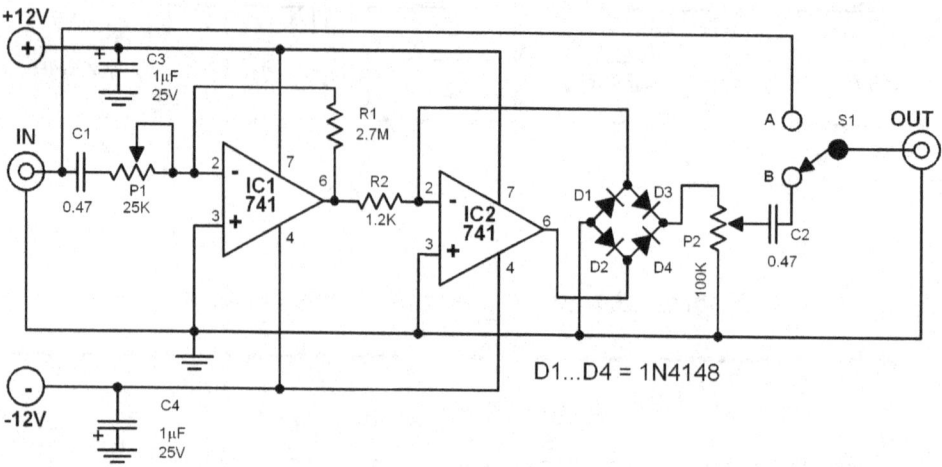

Diagram 4.0 Guitar Sound Effect

This sound effect circuit is actually a frequency doubler. It is commonly used by rock guitarists. It is also called octave shifter. It is one of the standards in a guitarist's arsenal of sound synthesizers and special effect devices. The guitar tone coming from the amplifier is shifted by one octave by this circuit.

Potentiometer P1 sets the input level of the signal and P2 sets the output amplitude of the amplified signal. The gain of the IC must be set through P1 to the point just short before the signal clips.

Diodes D1 to D4 function as bridge rectifier and doubles the signal frequency. Both diodes are connected in feedback to the amplifier so that their nonlinear character cannot affect the signal. Switch S2 is a bypass switch to turn off the circuit once a normal guitar tone is desired. This frequency doubler not only doubles the frequency of the signal but also changes its form. The output tone sounds synthetic compared to the original.

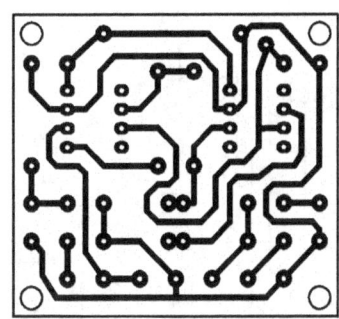

Figure 4.0 Printed Circuit Layout   **Figure 4.1** Parts Placement

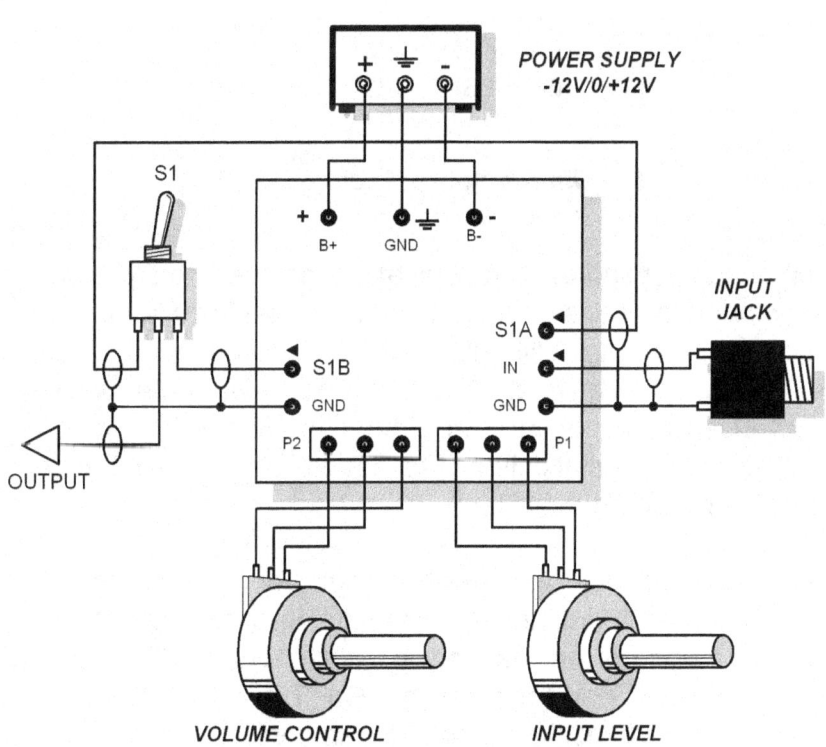

Figure 4.2 External Wirings

5 AUDIO SQUELCH

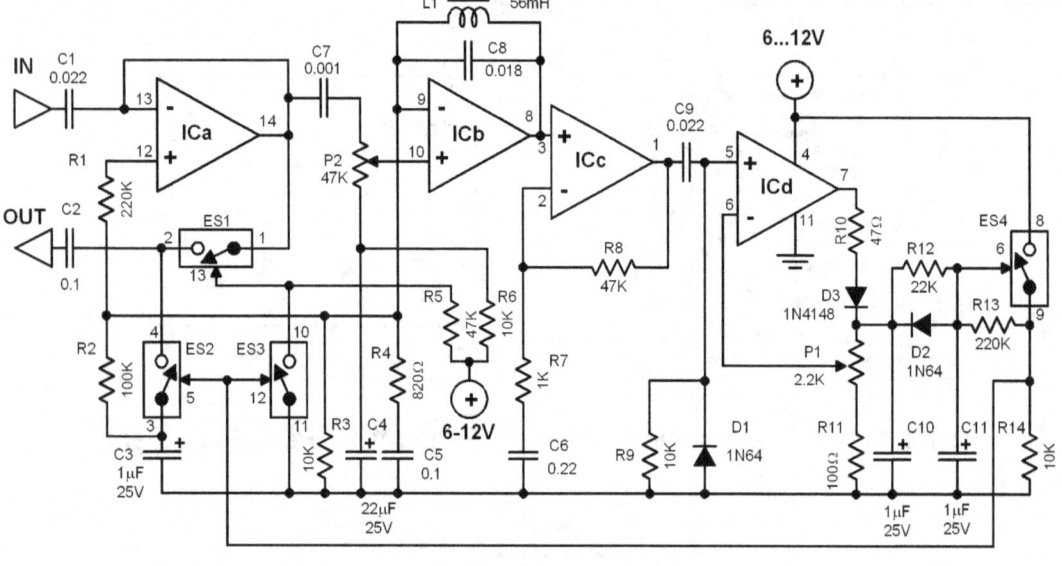

ICa...d = LM324
ES1...ES4 = 4066

Diagram 5.0   Audio Squelch

This squelch circuit is used in communication receivers to block out the audio (and noise) when no signal is being received. It works as a signal-to-noise ratio controlled squelch. It is originally designed for narrow band FM receivers. This circuit functions based on the fact that when the receiver detects no transmitted signal, it produces more noise than otherwise. When this noise exceeds a certain preset level, the circuit cuts off the connection between the demodulator and the audio amplifier input.

You can see in the block diagram how the circuit works. The output signal of the demodulator is no more connected to the input of the amplifier but instead enters the buffer ICa. The signal is then passed on to the bandpass filter ICb. The filtered signal is amplified by IC3 and rectified by ICd. The noise that is able to pass through ICb is amplified, rectified, and used as control signal for the electronic switch ES4. Switch ES4 in turn controls the switches ES1 and ES2.

When the noise level is below the threshold, ES1 is closed and ES2 is open so that the output of the demodulator is fed to the audio amplifier. Otherwise, when the noise level exceeds the threshold, ES1 opens and ES2 closes. The connection between the demodulator's output and the audio amplifier is cut off.

In using the circuit, the connection between the demodulator and the audio amplifier must be broken, and diverted to the proper terminals in the squelch circuit as shown in the Diagram 5.2. Potentiometer P2 adjusts the amplitude of the signal entering ICa. P3 sets the gain of rectifier ICd.

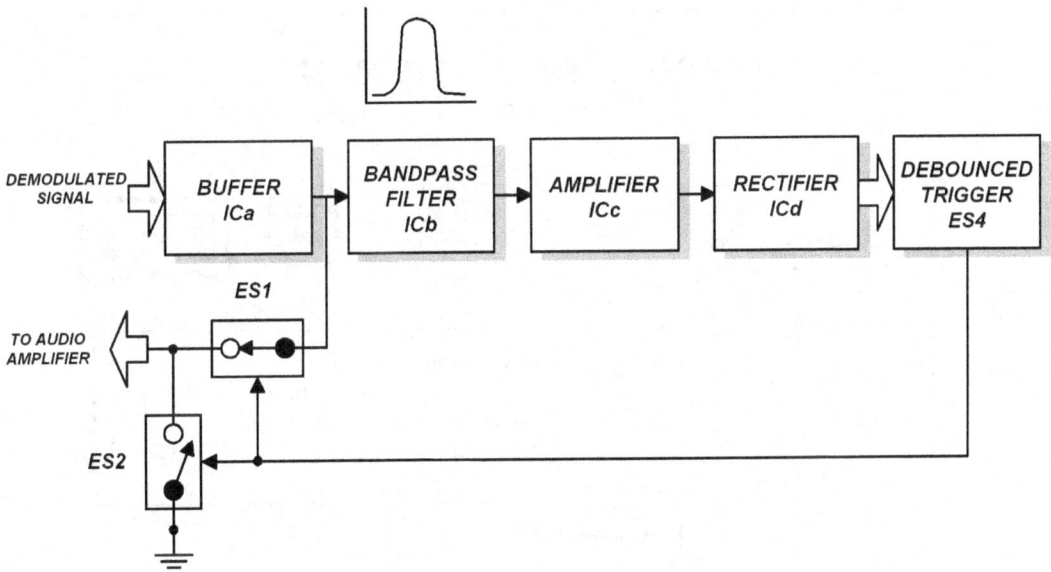

Diagram 5.1   Block Diagram of Audio Squelch

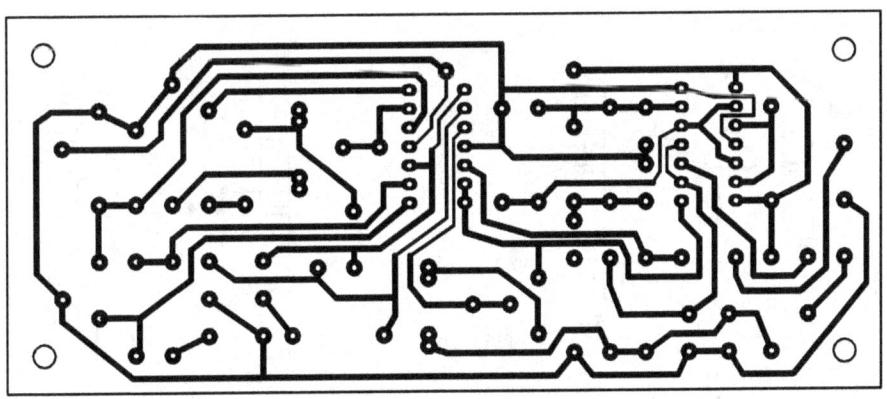

Figure 5.0   Printed Circuit Layout

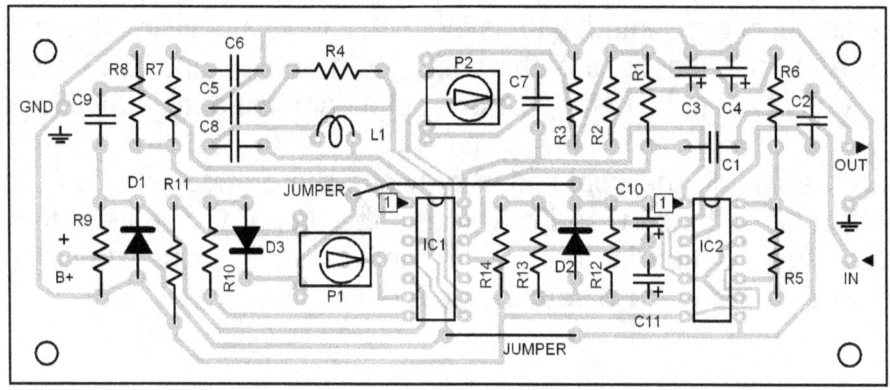

Figure 5.1 Parts Placement Layout

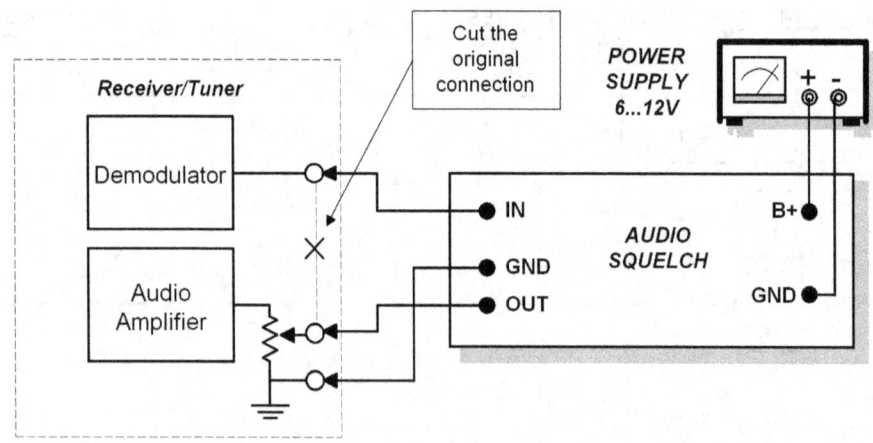

Diagram 5.2 Installation of Audio Squelch

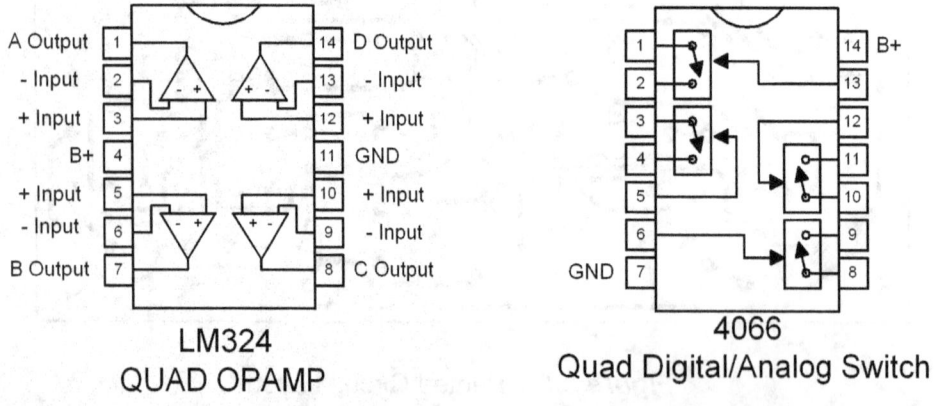

LM324
QUAD OPAMP

4066
Quad Digital/Analog Switch

6 HEATSINK THERMOMETER

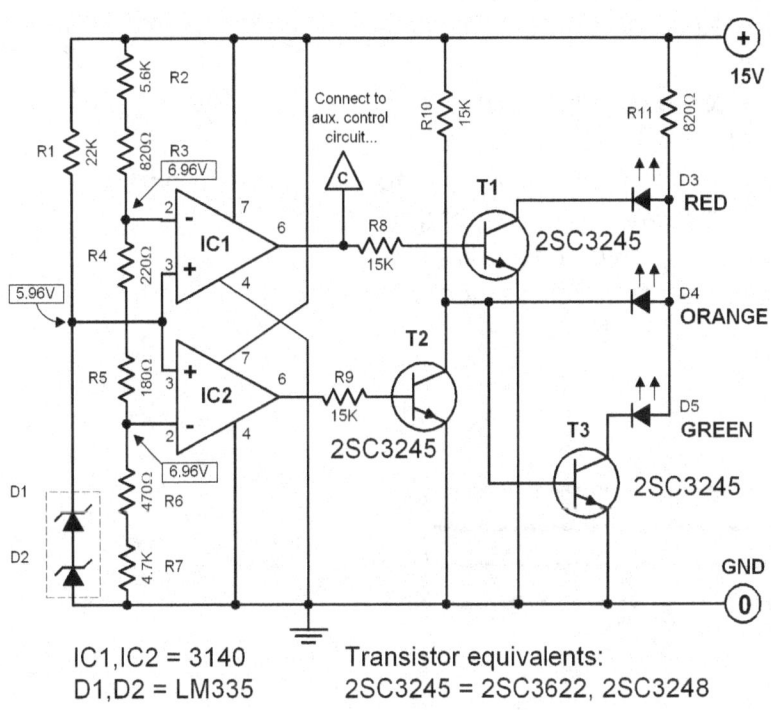

IC1,IC2 = 3140
D1,D2 = LM335

Transistor equivalents:
2SC3245 = 2SC3622, 2SC3248

Diagram 6.0 Heatsink Thermometer

This "thermometer" monitors the temperature of a heatsink, and displays its approximate level through three LEDs. The green LED lights up as long as the temperature is less than 50°C. The orange LED lights up when the temperature is between 50°C and 75°C. The red LED lights up when the temperature rises above 75°C.

The temperature sensor is made of two special zener diodes. The zener voltage is exactly 5.96 V when the temperature is 25°C. This zener voltage increases by 20 mV per degree centigrade.

Sensors D1 and D2 must be installed on the heat sink as close as possible to the final transistors to minimize thermal losses.

Of course, you can use this temperature sensor for purposes other than blinking LEDs. One of the best applications of this circuit is the automatic control of an air blower that cools the heatsink down to a safe temperature level. Connect a relay to transistor T4. The relay, in turn, controls any kind of device connected to it like, for example an air blower. Never connect the device directly to transistor T4.

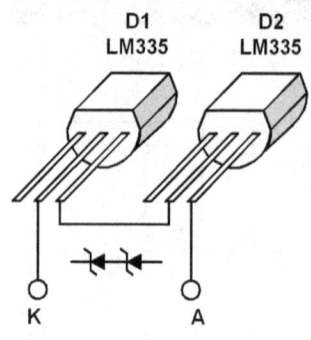

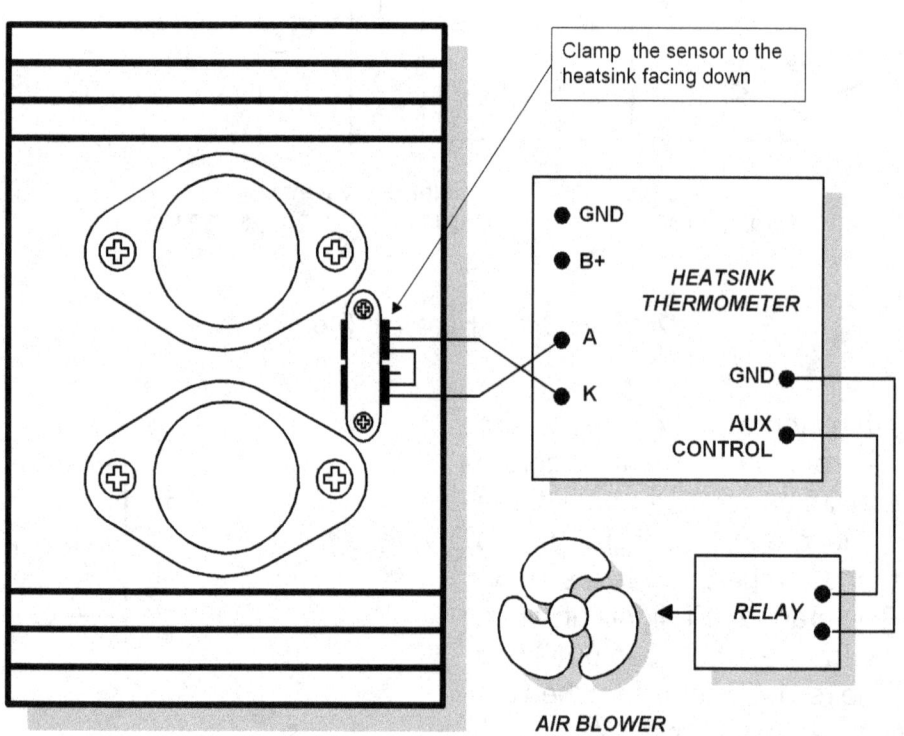

Figure 6.0 External Wirings

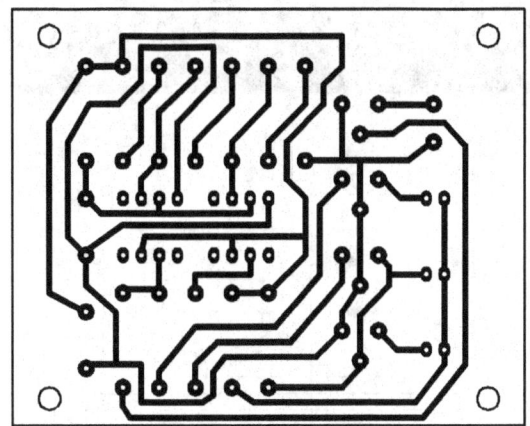

Figure 6.1   Printed Circuit Layout

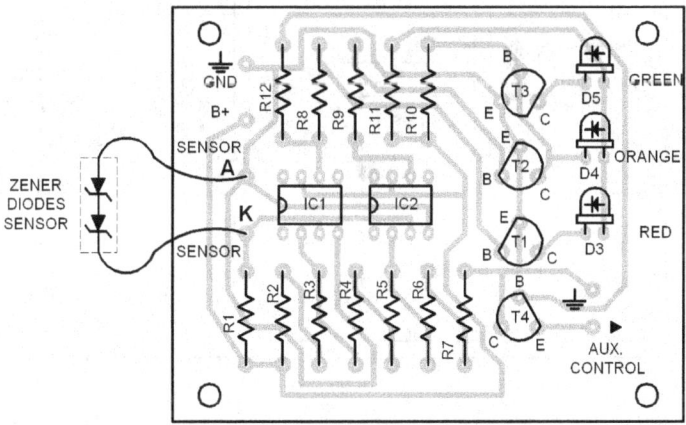

Figure 6.2   Parts Placement Layout

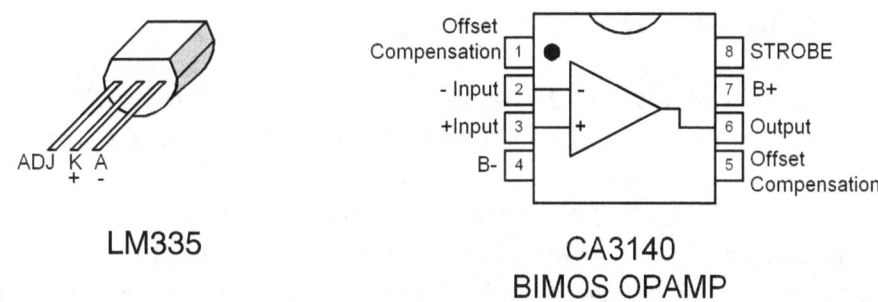

LM335

CA3140
BIMOS OPAMP

7 SPEAKER PROTECTOR

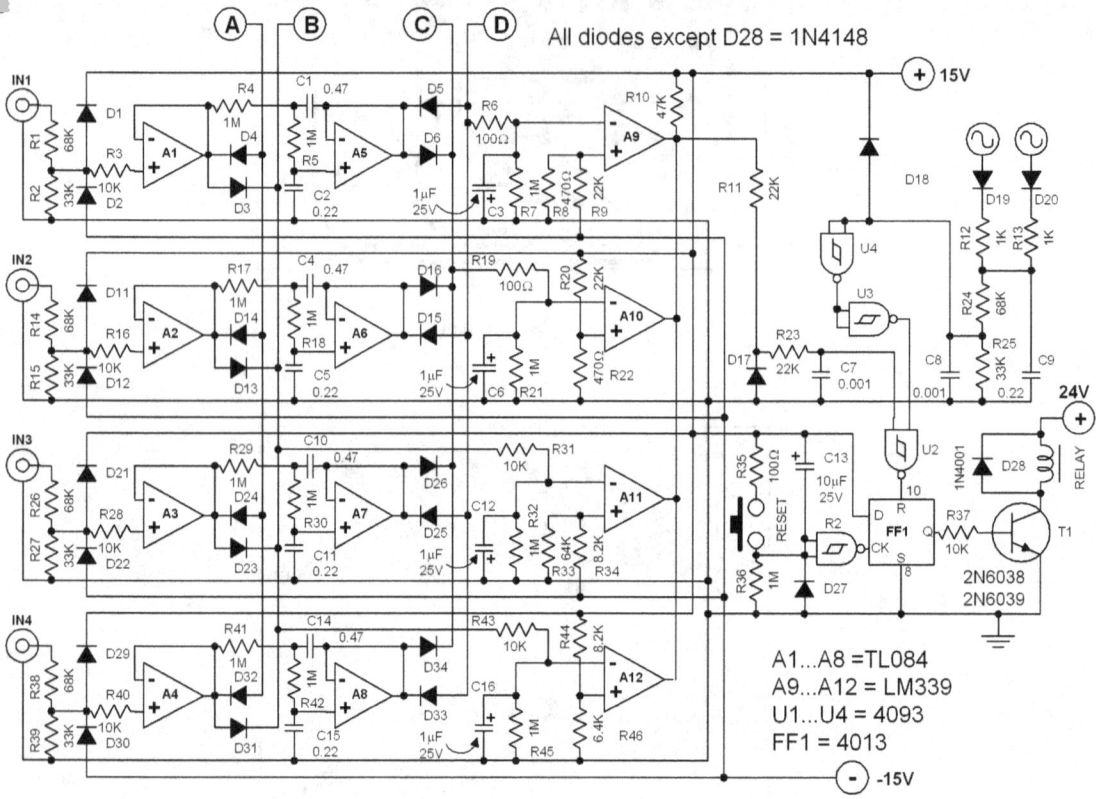

Diagram 7.0 Speaker Protector

This circuit protects the speaker when the final amplifier malfunctions. Conversely, it protects the final amplifier when the speaker overloads it. The protector circuit is originally designed for active-box builders who often encounter malfunctions in the modules they are experimenting.

The circuit is composed of three stages: the input buffers, the lowpass filter and the relay stage. The input stage can monitor four signals. The number of inputs can be readily expanded according to actual needs. The relay will activate when a high input signal comes out of the final amplifier or when the speaker overloads the final stages. The speaker is connected to the contact terminals of the relay so that the speaker will be automatically disconnected when a malfunction occurs. S1 resets the relay.

8 STEREO INDICATOR

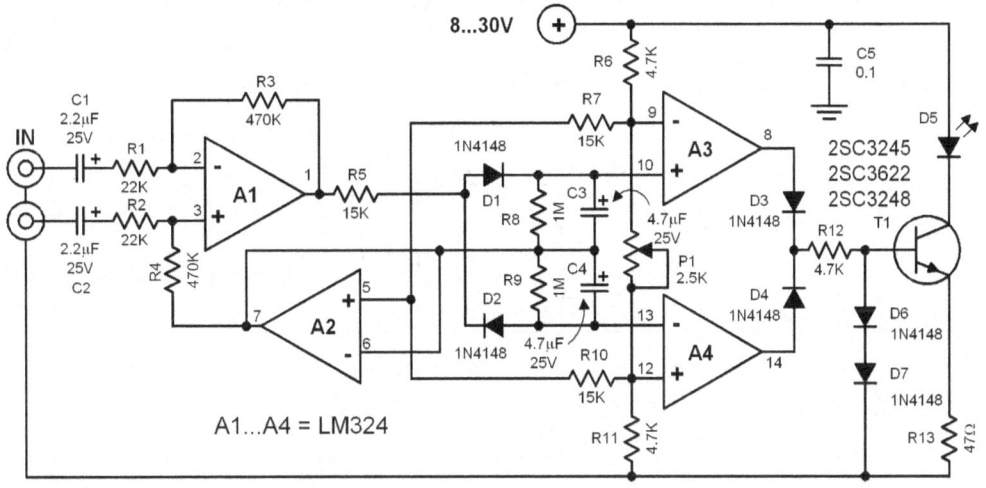

Diagram 8.0 Stereo Indicator

This circuit indicates whether the signal being received by your FM radio is a true stereo signal. The principle applied is very simple: the left and right channels are compared. When a difference between the two channels appears, the signal is stereo. The circuit must be installed inside the receiver and connected before the balance and volume controls of the receiver. The sensitivity can be adjusted by P1.

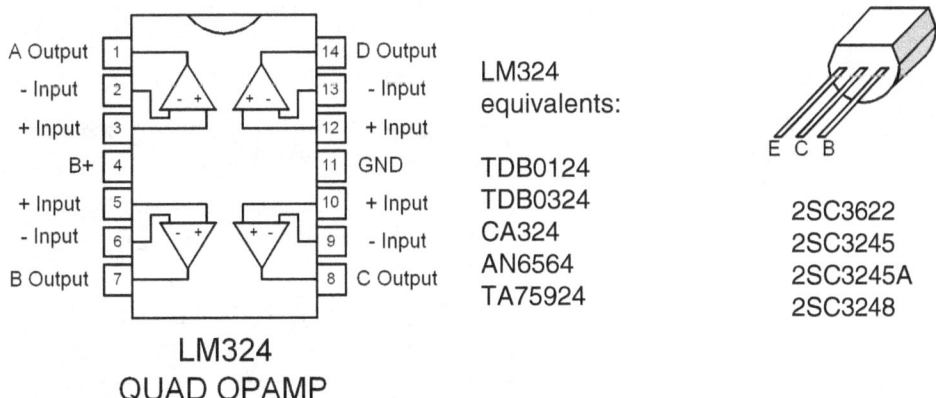

LM324
QUAD OPAMP

LM324
equivalents:

TDB0124
TDB0324
CA324
AN6564
TA75924

2SC3622
2SC3245
2SC3245A
2SC3248

9 VIDEO AMPLIFIER

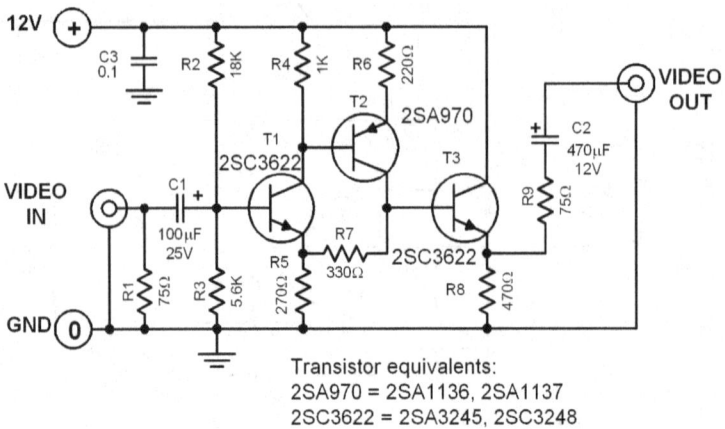

Diagram 9.0 Video Amplifier

This is a universal video amplifier which is very simple to construct and which does not need special components. By connecting this circuit in a long coaxial cable it compensates for the signal losses within the cable. The circuit is composed of 2-stage amplifier with an emitter follower as impedance converter.

 The power supply must be stabilized properly to avoid signal interferences.

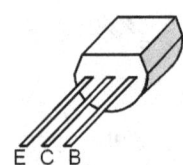

E C B

2SA970	2SC3622
2SA1136	2SC3245
2SA1137	2SC3248

Technical Data	
Amplification factor=	approx. 2
Bandwidth=	over 20 MHz
Supply voltage=	12 V
Current consumption=	20 mA

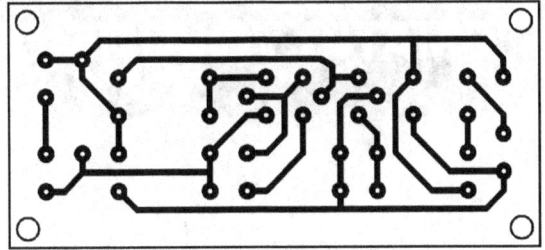

Figure 9.0 Printed Circuit Layout

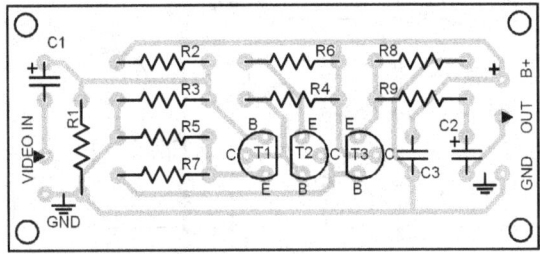

Figure 9.1 Parts Placement

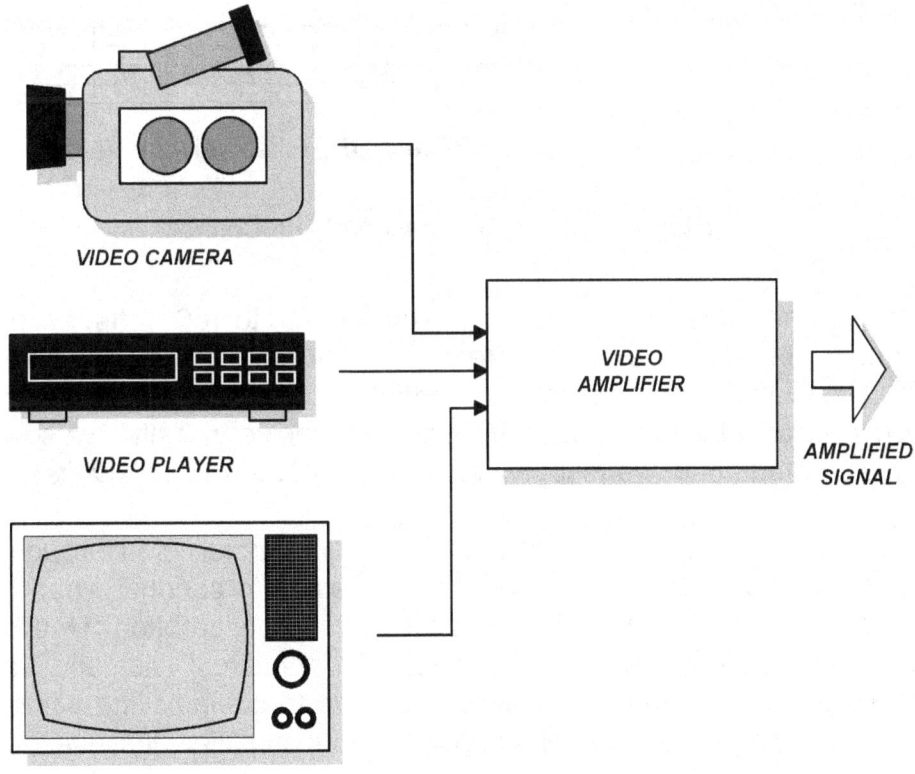

10 AUTOMATIC VOLUME CONTROL

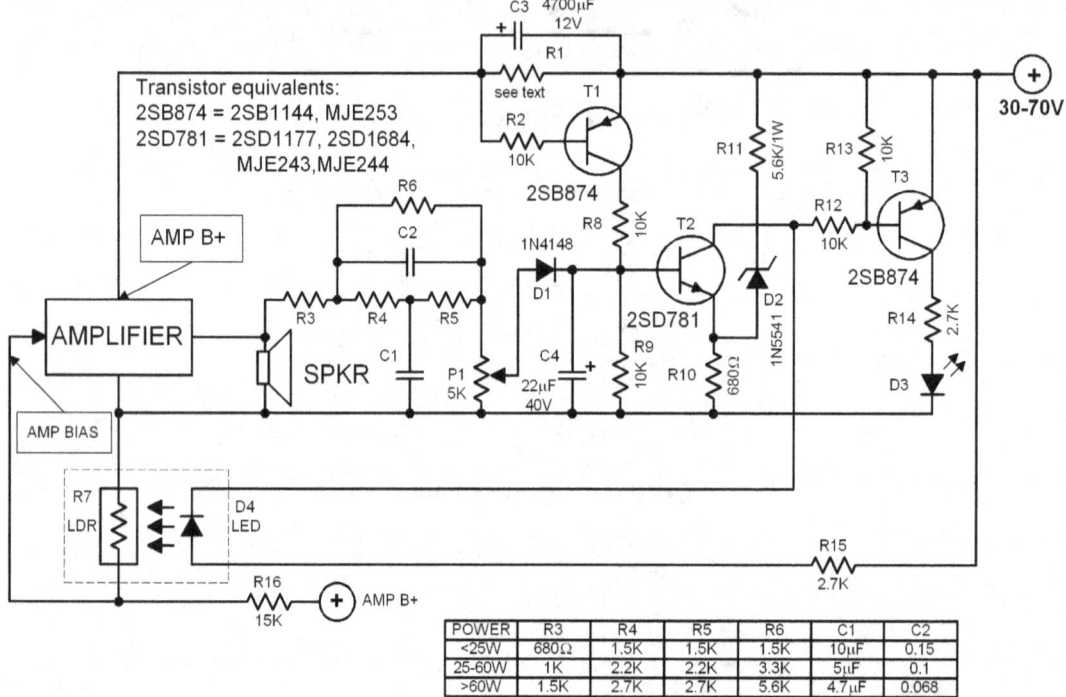

Table 10.0 Filter Network Values

POWER	R3	R4	R5	R6	C1	C2
<25W	680Ω	1.5K	1.5K	1.5K	10µF	0.15
25-60W	1K	2.2K	2.2K	3.3K	5µF	0.1
>60W	1.5K	2.7K	2.7K	5.6K	4.7µF	0.068

Diagram 10.0 Automatic Volume Control

This circuit limits the power output of an amplifier (10 to 100 watts) according to the combination of its output level, frequency, current consumption and supply voltage. The values are of course dependent on the actual amplifier used. This circuit can be adapted to the individual amplifier type very easily. It also accepts a very wide range of power supply voltages from 30 volts to 70 volts!

For example the current consumption of the amplifier is sampled through R1. As you know very well, the silicon transistor T1 starts to conduct when its base voltage reaches around 0.56 volts. Once the current consumption of the amplifier produces a voltage drop of 0.56V at resistor R1, the circuit (actually transistor R1) starts to limit the amplifier's gain. You have to compute the value of R1 yourself basing on the current consumption of your amplifier. Just use the Ohm's law.

The actual gain control of the amplifier is done by the LDR/LED combination shown in the circuit. The LDR/LED combination must be housed inside a light-proofed box. The AVC is also connected to the loudspeaker or to the output terminals of the final amplifier. A filter network is connected right after the speaker.

The sample signal passes first through the filter before it is processed by the main circuit. The component values are dependent on the frequency and output power. The component values for the filter are listed in Table 10.0. The limit threshold can be set through the trimmer P1. If you want to detect the signal peaks only, remove the filter circuit. In such case the circuit is frequency independent.

E C B

2SB874	2SB1144
2SD781	2SD1177
2SD1685	MJE243
MJE244	MJE253

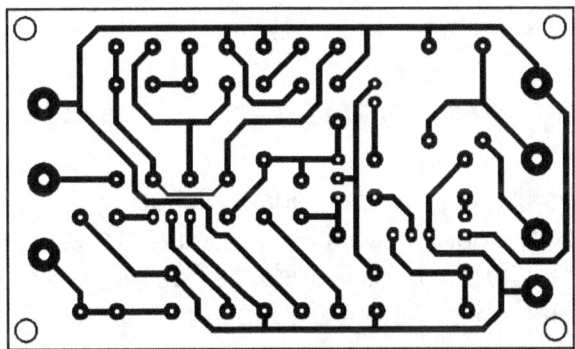

Figure 10.0 Printed Circuit Layout

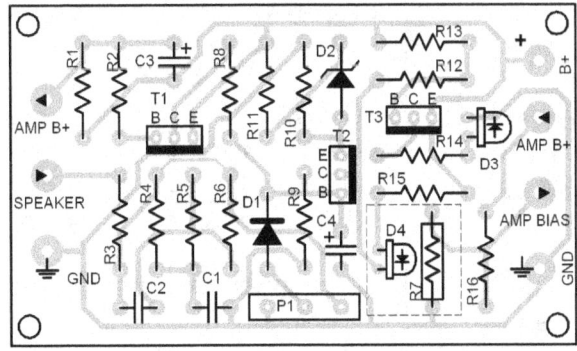

Figure 10.1 Parts Placement

11 CASSETTE PREAMP

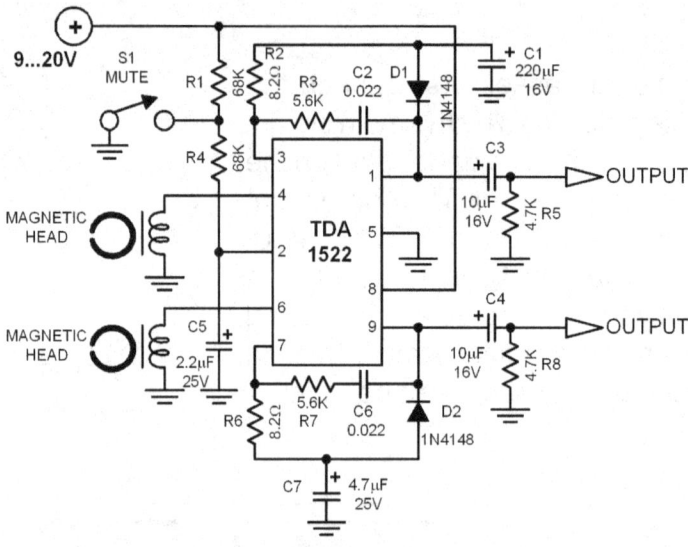

Diagram 11.0 Cassette Preamp

This circuit amplifies the very weak signals picked up by a cassette magnetic head. This preamp is highly insensitive to noise. The usual klick or popping sound at the start of every playback is absent since the preamp inputs are directly connected to the head coils without a coupling capacitor.

Parts List
R1,R4= 4.7K
R2,R3= 68K
R5,R6= 5.6K
R7,R8= 8.2 ohms
C1,C2= 10µF/10V
C3= 2.2µF/25V
C4,C5= 22µF
C6,C7= 220µF/10V
D1,D2= 1N4148
IC1= TDA 1522

Technical Data	
Supply Voltage=	7.5V - 23 V(typical 8.5V)
Standby amplification=	90 dB
Current consumption=	7.5 mA
Output power =	1.5 Vrms maximum
Distortion factor=	0.05%
Input resistance=	over 200K
Output resistance=	less 1K

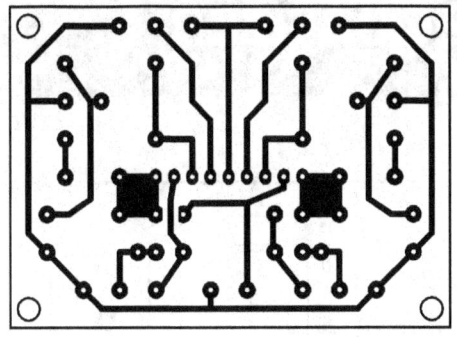

Figure 11.0 Printed Circuit Layout

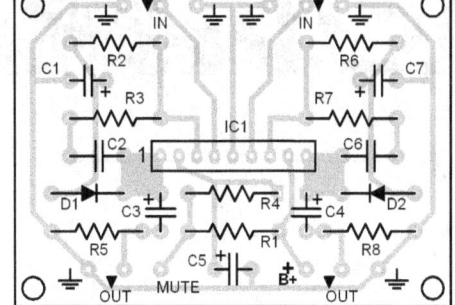

Figure 11.1 Parts Placement

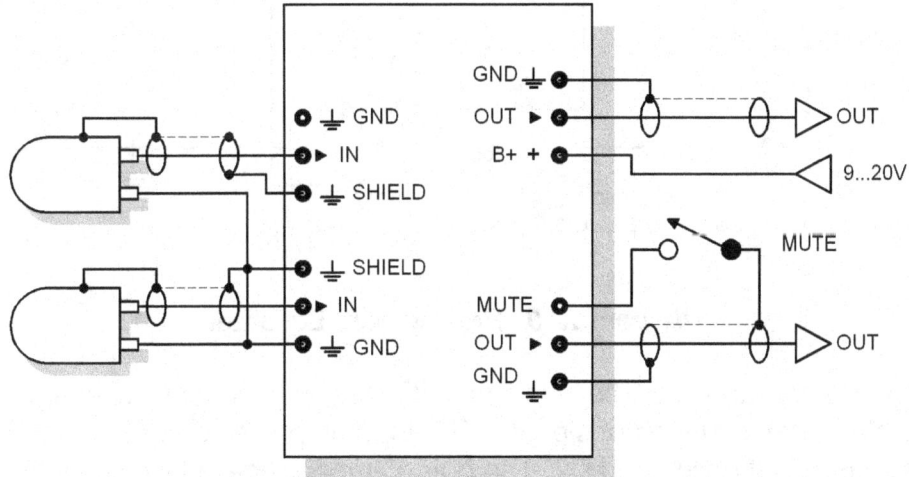

Figure 11.2 External Wiring for 2 Mono Heads

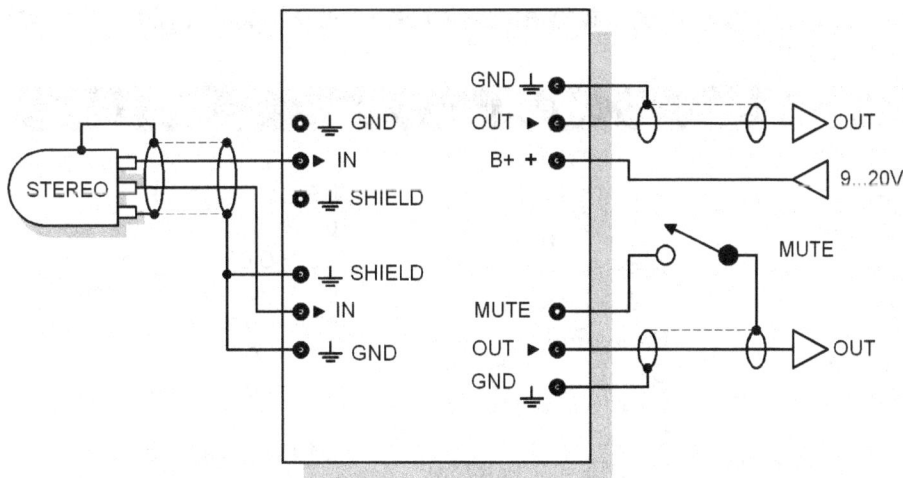

Figure 11.3 External Wiring for single Stereo Head

12 PREAMP w/ EQUALIZER

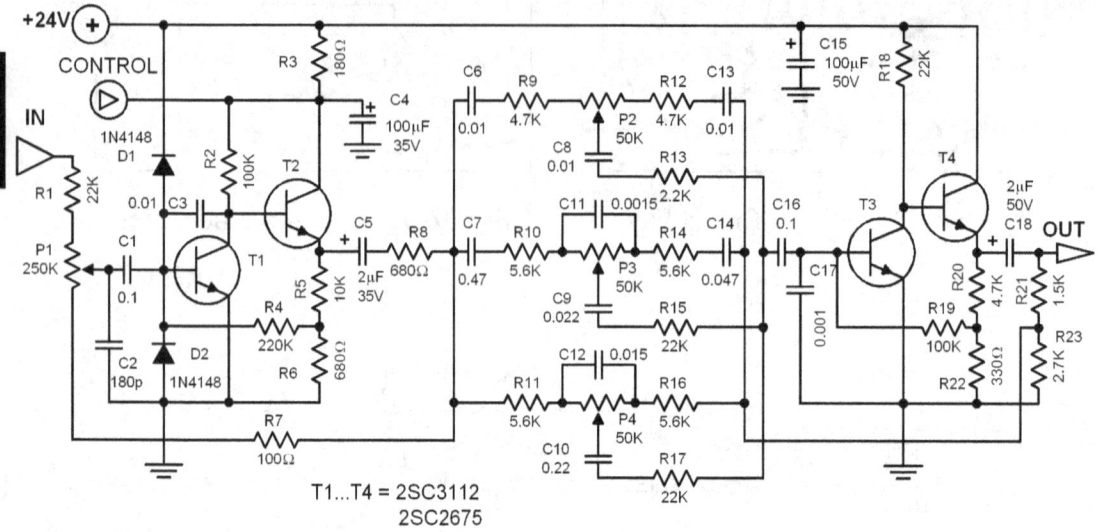

Diagram 12.0 Preamp with Equalizer

This universal preamp can be used for different applications. It enables the amplification of many types of signals. The input is protected from overdrive by the diode combination D1 and D2. The output impedance is low which enables it to be connected to any type of final amplifier without problems. The preamp is composed of two dc-coupled amplifier stages. A simple three-way equalizer is connected between these two amplifier stages. The amplification level is controlled by P1.

Parts List

Resistors:
R1,R15,R17,R18= 22K
R2,R19= 100K
R3= 180Ω
R4= 220K
R5= 10K
R6,R8= 680Ω
R7= 100W
R9,R12,R20 = 4.7K
R10,R11,R14,R16= 5.6K

R13 = 2.2K
R21= 1.5K
R22= 330Ω
R23= 2.7K

Capacitors:
C1,C16= 0.1/50V ceramic
C2= 180p/50V ceramic
C3,C6,C8,C13= 0.01/50v
C4,C15 = 100µF/35V

C5,C18= 2µF/35V
C7 = 0.47/50V ceramic
C9 = 0.022/50V ceramic
C10 = 0.22/50V ceramic
C11 = 0.0015/50V ceramic
C12 = 0.015/50V ceramic
C14 = 0.047/50V ceramic
C17= 0.001/50V ceramic
D1,D2= 1N4148
T1,T2,T3,T4 =2SC3112(2SC2675)

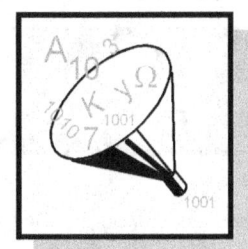

HOBBY & SHOP

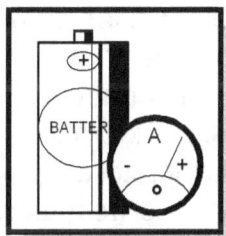

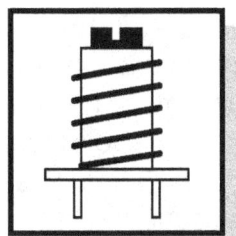

13 FM TRANSMITTER

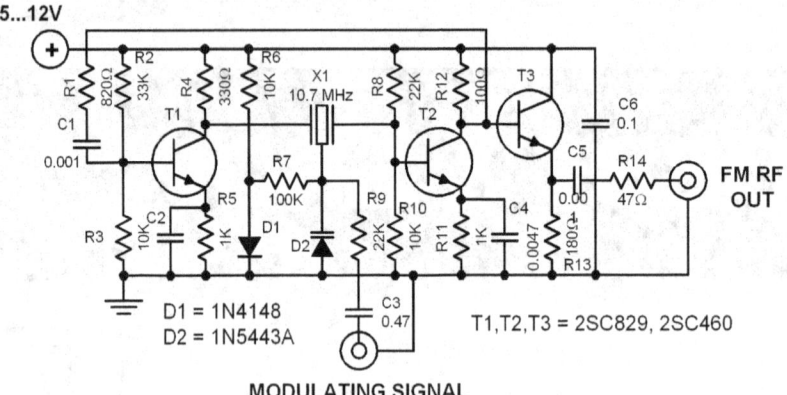

Diagram 13.0 FM Transmitter

This FM transmitter uses a standard ceramic filter commonly used in color television receivers to filter color information. The necessary RF signal is produced by the astable multi-vibrator coupled to the ceramic filter. The frequency is dependent on the filter's frequency. The phase shift of the ceramic filter is zero degree. This transmitter can be audio modulated. The oscillator modulation is controlled by the varicap diode D2. It has a modulation bandwidth of 13kHz.

This transmitter is designed to calibrate FM tuners and to serve as signal generator for calibrating VHF amplifiers. They can also be used as mini broadcast transmitters. Figure 13.2 shows some possible applications of the circuit.

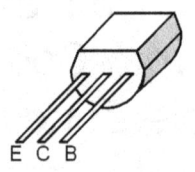

2SC460
2SC829

Parts List	
Transistors:	Capacitors(ceramic):
R1 = 820Ω	C1,C5= 0.001/50V
R2 = 33K	C2,C4= 0.0047/50V
R3,R6,R10,= 10K	C3= 0.47/50V
R4 = 330Ω	C6 = 0.1/50V
R5,R11,R13 = 1K	D1 = 1N4148
R7 = 100K	D2 = 1N5443A
R8,R9 = 22K	Transistors:
R12 = 100Ω	T1,T2,T3 = 2SC829
R14 = 47Ω	(2SC460)

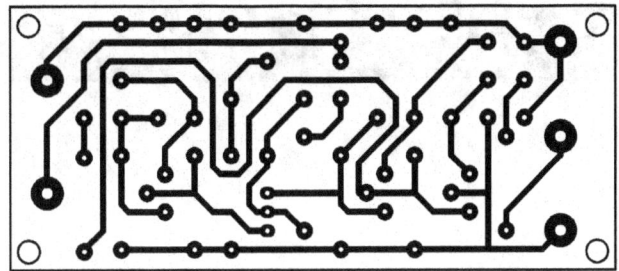

Figure 13.0 Printed Circuit Layout

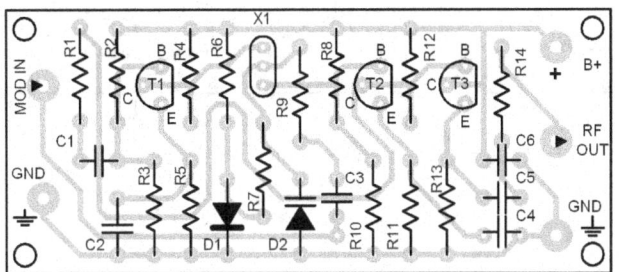

Figure 13.1 Parts Placement Layout

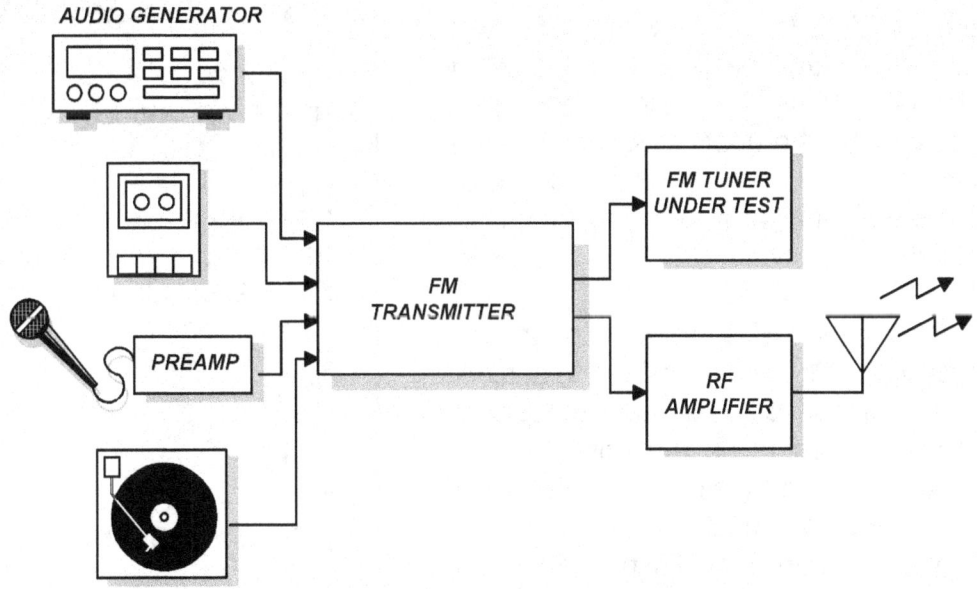

Figure 13.2 Sample applications

14 LED OPTO-COUPLER

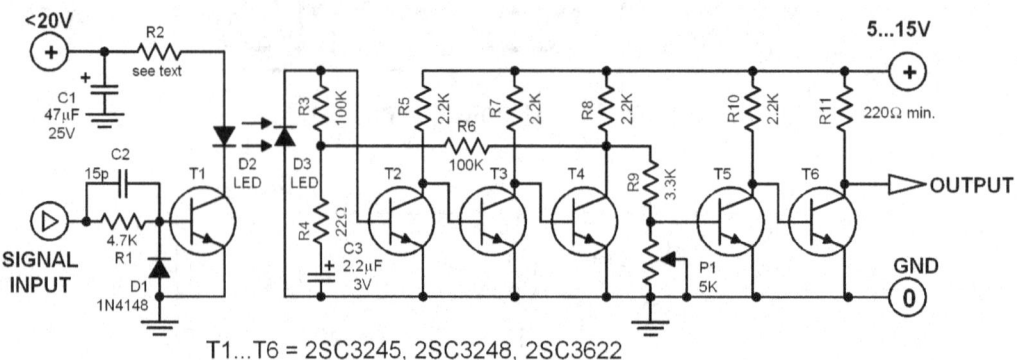

T1...T6 = 2SC3245, 2SC3248, 2SC3622

Diagram 14.0 LED Opto-Coupler

This opto-coupler circuit uses two ordinary LEDs as the opto-coupler element. When a voltage of 2V is fed to the input, LED2 will generate a voltage that is (after amplified by the following transistors) strong enough to switch on T5. The upper frequency limit is around 38 kHz. Very low pulse frequencies on the other hand is almost meaningless for the circuit.

Eventhough the LEDs are quite insensitive to stray light, it is still a good idea to enclose them with an opaque box. Resistor R2 must be selected basing on the power supply voltage. It can be between 680 ohms and 1K. Resistor R11 must not be lower than 220 ohms. Transistor T6 has an open-collector character.

Parts List

R1 = 4.7K
R2 = 680Ω - 1K(see text)
R3,R6 = 100K
R4 = 22Ω
R5,R7,R8,R10 = 2.2K
R9 = 3.3K
R11 = 220Ω minimum
P1 = 5K trimmer
C1 = 47µF/25V
C2 =   15p/50V
C3 = 2.2µF/3V
D1,D2 = LED
T1...T6 = 2SC3622
(2SC3245)(2SC3248)

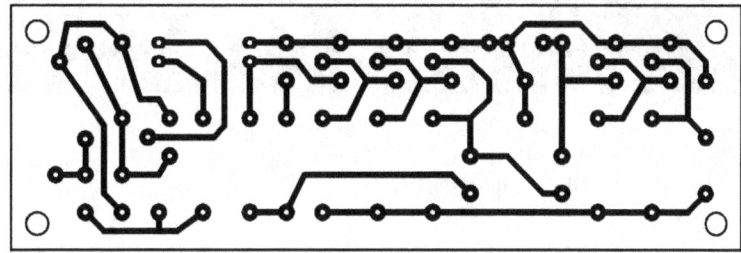

Figure 14.0 Printed Circuit Layout

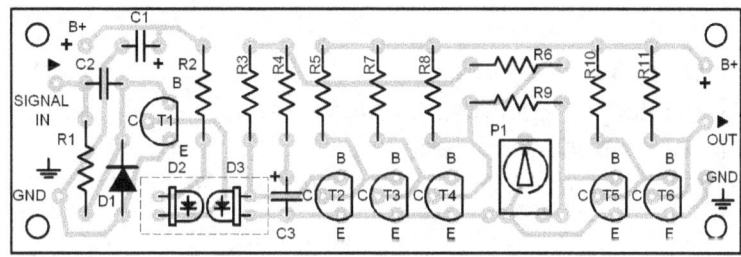

Figure 14.1 Parts Placement Layout

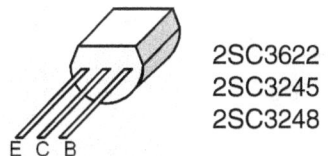

2SC3622
2SC3245
2SC3248

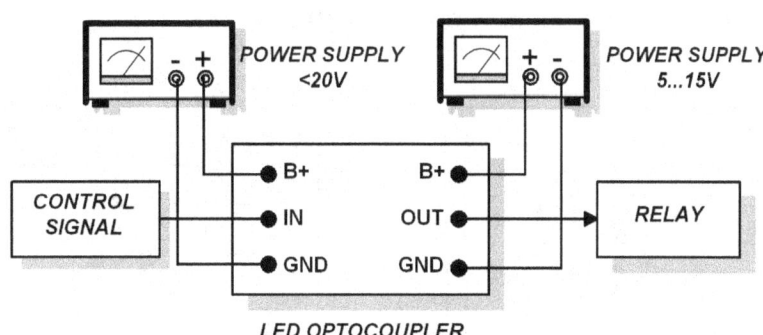

Figure 14.2 External Wirings

15 VARIABLE ZENER DIODE

Zener diodes are excellent components for regulating voltages. However, they are available only in fixed voltage ratings. When you want to regulate a voltage level which is different from any of the standard zener voltages, a problem arises.

Actually, it is never a problem if you use this circuit. The voltage regulated by this circuit can be varied. The zener voltage(V_Z) of the circuit can be found with:

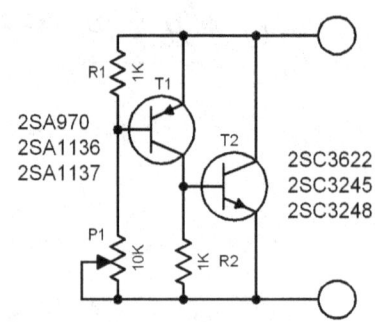

Diagram 15.0 Variable Zener Diode

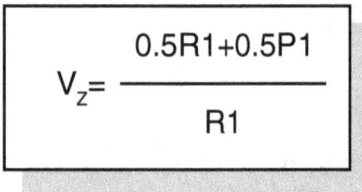

$$V_Z = \frac{0.5R1 + 0.5P1}{R1}$$

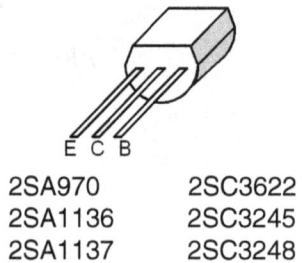

2SA970	2SC3622
2SA1136	2SC3245
2SA1137	2SC3248

This circuit design is ideal for zener voltages above 0.8 Volts. The dynamic internal resistance remains less than 3 ohms. The circuit offers the advantage of having very low internal resistance specially in the range from 1V to 3V. The zener voltage can be adjusted through trimmer P1.

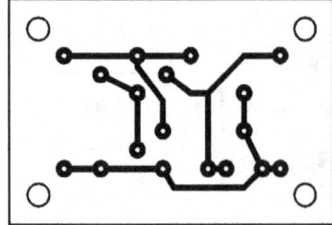

Figure 15.0 Printed Circuit Layout

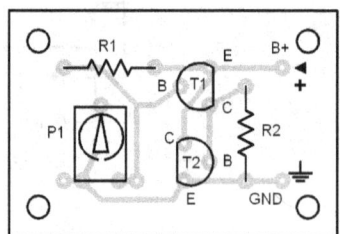

Figure 15.1 Parts Placement

16 LED REFERENCE DIODE

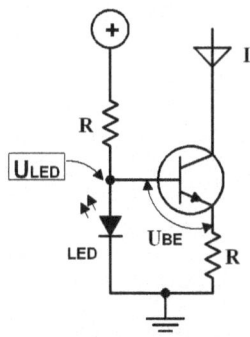

Diagram 16.0
LED Reference Diode

The voltage drop in a light emitting diode is between 1.4V and 2.2 V when a current of around 5 to 10 mA flows through it. The voltage drop is independent of the LED's type. A LED's temperature coefficient is -1.5mV/°C. That means, when the temperature rises by about 1°C the voltage drop decreases by around 1.5mV.

This characteristic is exploited in this circuit to create a constant current source which is almost independent from temperature.

The collector current can be found with:

 Take note that the value of U_{LED} varies slightly from one type of LED to another.

$$I_z = \frac{U_{LED} - U_{BE}}{R}$$

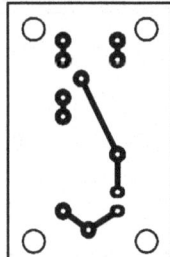

Figure 16.0
Printed Circuit Layout

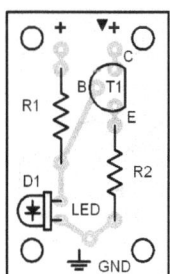

Figure 16.1
Parts Placement

17 SELECTIVE FILTER

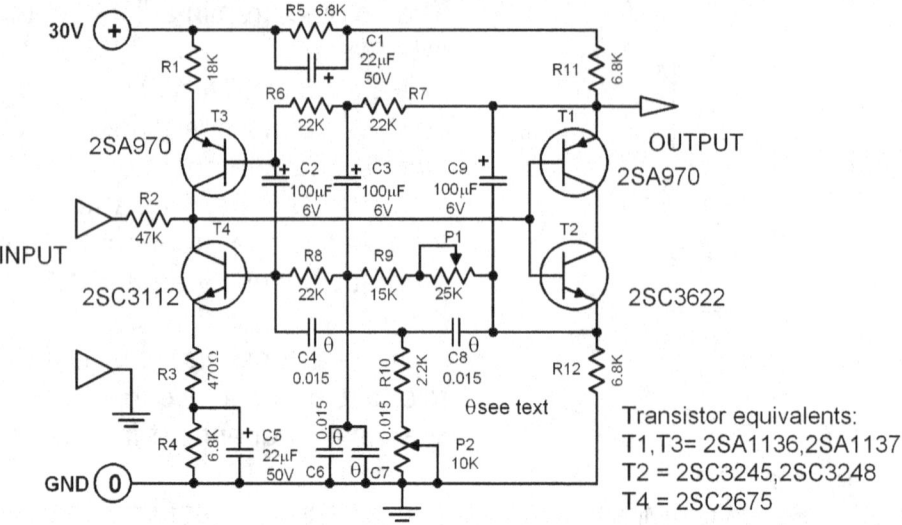

Diagram 17.0 Selective Filter

This circuit is an audio filter with a so-called double T network. It is a narrow bandpass filter. It can be set at a certain frequency (fo) with trimmer potentiometers. All frequencies other than the fo are attenuated. The circuit has a complementary emitter follower composed of T1 and T2 which are controlled through R1. The emitter of T1 or T2 is the output. The output is connected to a balance amplifier T3/T4 through the double T network of P1,P2,R6 to R10, and C4,C6,C7,C8 (marked in the diagram). The amplification(A) of the balance amplifier is:

$$A = \frac{2R2}{R1} = \frac{2R2}{R3}$$

$$fo = \frac{1}{RC}$$

$$R6, R7, R8 = 2\mathbf{R}$$

$$R9 + P1 = 2\mathbf{R}$$

$$R10 + P2 = \mathbf{R}/2$$

Potentiometers P1 and P2 set the maximum output at the frequency fo. When **R** = 11K and C4,C6,C7,C8= 0.015 , the fo is around 1000 Hz (1kHz).

18 ELECTRONIC PUSHBUTTON

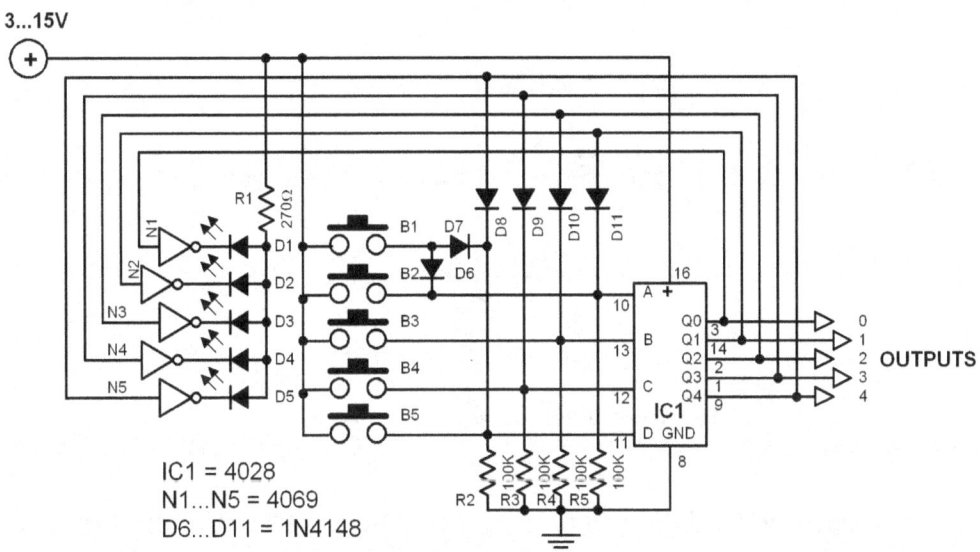

Diagram 18.0 Electronic Pushbutton

The ideal multiple contact switch that is designed (and must function) as break-before-contact selector must really do what it is supposed to do- that is, to break the old contact before making a new contact. This is, however, not guaranteed in mechanical switches. You can just imagine what happens when two or more contacts are simultaneously connected with each other.

Here comes the purely electronic "break-before-contact" switch to the rescue. The advantage of this switch in comparison to its mechanical equivalent is: when more than one pushbutton are pressed simultaneously, all "contacts" remain open. The switch will only react when only one button is pressed at one time. Additionally, a LED shows which button was pressed and which line is currently active. Button1 is a normal pushbutton (non-latching) which opens the contact once it is released. Buttons B2 up to B5 are latching. A contact remains closed until another button is pressed.

19 INFRARED DETECTOR

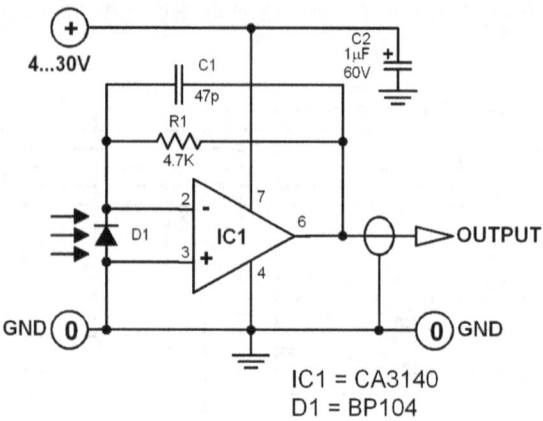

Diagram 19.0   Infrared Detector

This circuit is usually used as sensor for infrared activated switches. It is applied for burglar alarm circuits, fire alarms, temperature regulators, etc. It can also be used to automatically control light intensity. This circuit was originally used to control the brightness of a 7-segment display in a commercial aircraft.

The circuit is composed of a photodiode in a short-circuited configuration and an opamp. The opamp's gain is set by resistor R1. You can also experiment with this infrared detector. Connect it to an appropriate audio amplifier and listen to the unusual sound of continously changing heat in your living room or kitchen.

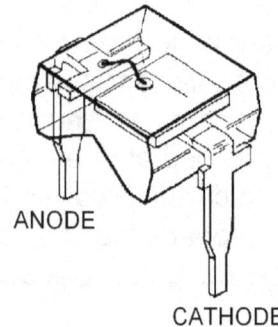

ANODE

CATHODE

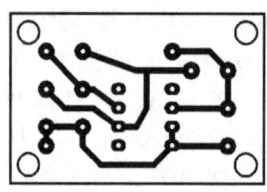

Figure 19.0
Printed Circuit Layout

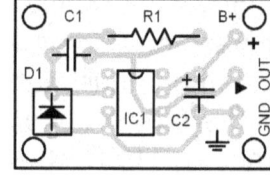

Figure 19.1
Parts Placement

20 DUMMY CAR ALARM

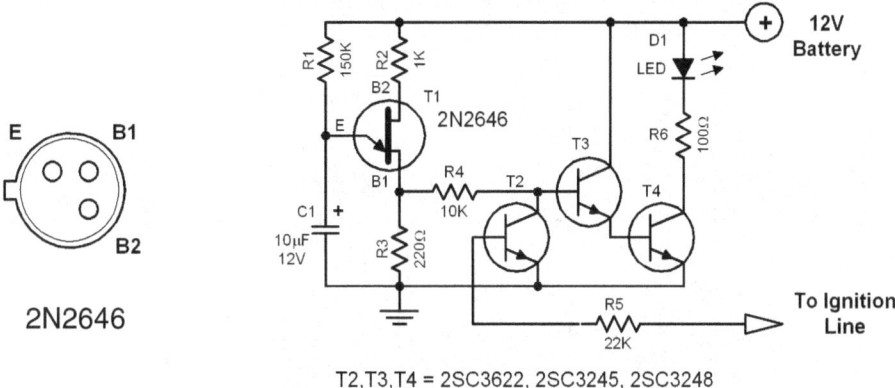

2N2646

T2,T3,T4 = 2SC3622, 2SC3245, 2SC3248

Diagram 20.0   Dummy Car Alarm

This circuit drives a LED to blink. When constructed inside a box it will simulate an alarm circuit and fools a car thief into thinking that the car is "guarded". Of course, this will only make sense if you don' forget to activate the "alarm". Now, since humans are well known for forgetting to turn alarms on, this circuit activates itself automatically. It functions only when the car is not running and the ignition key is turned off (usually when parked).

The circuit is designed so that the current consumption is very low - around 2 mA. Although the LED lights bright enough to serve its purpose, its consumes very little current - around 2 mA. This is achieved by pulsing the LED. Since the pulsing frequency is very high, the LED appears to light continously.

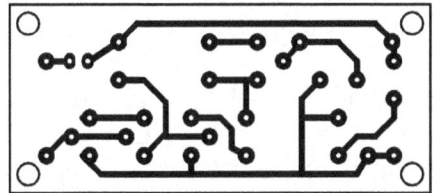

Figure 20.0 Printed Circuit Layout

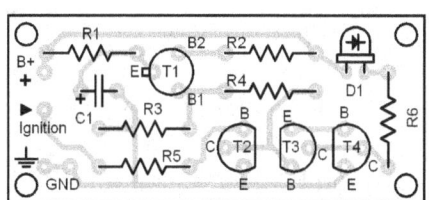

Figure 20.1 Parts Placement

21 VERSATILE TIMER

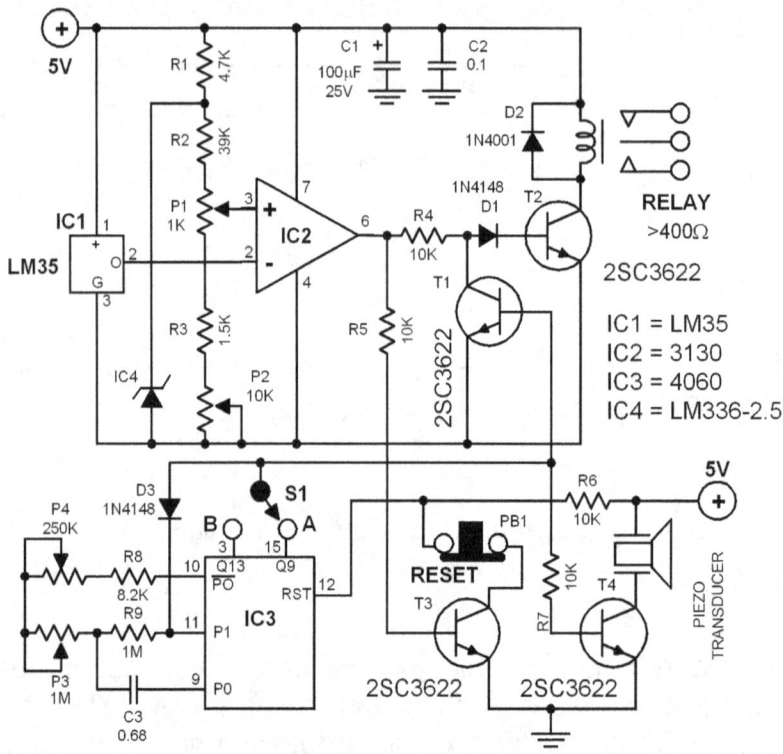

Diagram 21.0 Versatile Timer

This circuit switches a relay in response to a temperature value or elapsed time. The temperature sensor of the circuit uses an LM35 IC that gives an output voltage of 10mV/°C. This voltage is monitored by the comparator IC2. The reference level of IC2 is set by P1 and P2. T1 conducts as long as the temperature is below the threshold value. The timer part is made up of IC3. The timer delay range can be selected by S1 from 6...90 seconds and 1.5s...25 minutes. Once the timer switches on, T2 blocks and the relay is disengaged. The timer can be reset through S2 but only if the maximum temperature is not reached yet.

Calibration: Us a DVM and test the junction of R3 and P1, while adjusting P2 until the voltmeter shows 100mV. That means a reference temperature of 10°C. Test pin3 of IC3 while adjusting P1 so that the desired "switch off" temperature is set. Then set P3 to minimum position, and switch S1 to position A. Set P4 so that the relay will switch off 5...6 seconds after the reset button is pressed. The desired delay time can be set through P3. **The relay coil must have a resistance of not less than 400 ohms.**

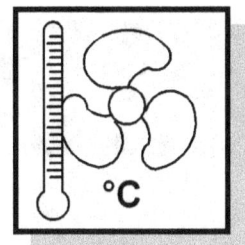

HOUSE & CAR

22 HOUSEPHONE AMPLIFIER

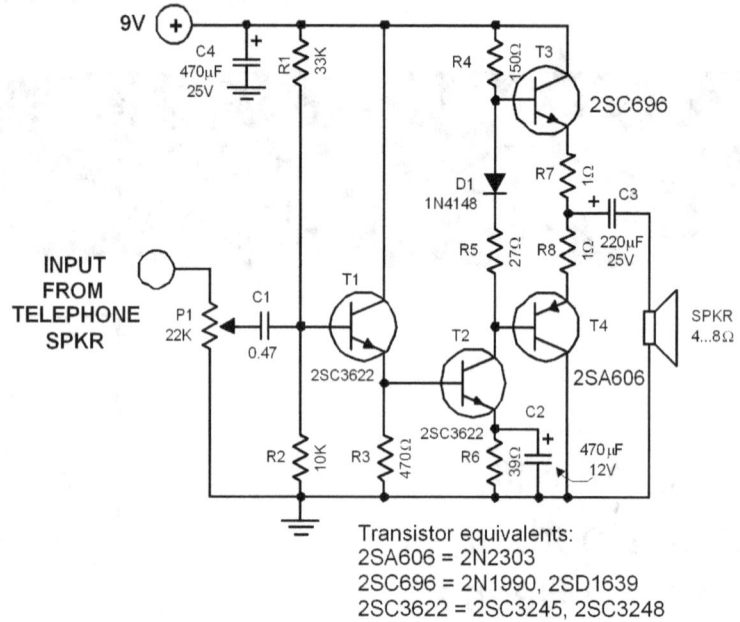

Diagram 22.0 Housephone Amplifier

This housephone (or telephone) amplifier doesn't need an inductive pickup. It is directly connected to the phone's speaker capsule. This way, interference is avoided.

The input's polarity is not important since the circuit has its own power supply. Potentiometer P1 sets the volume of the audio signal. The standby current consumption is around 30 mA. If you want to achieve better sound quality, replace resistor R5 with a 100W trimmer and set the standby current of T3 & T4 to 10 mA.

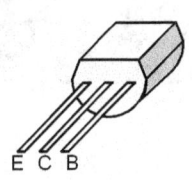

2SC3622
2SC3245
2SC3248

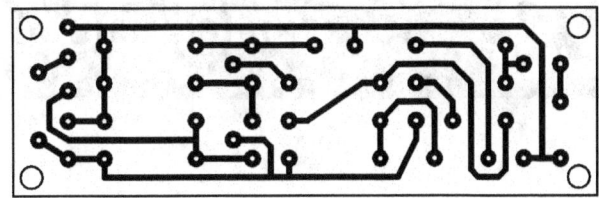

Figure 22.1 Printed Circuit Layout

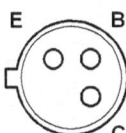

2SA606   2SC696
2N2303   2SD1639
2N1990

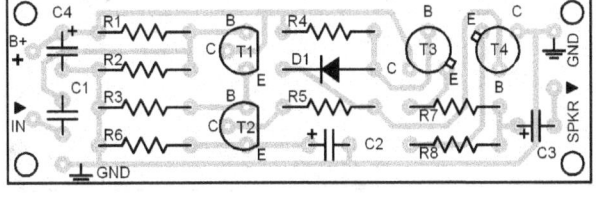

Figure 22.0 Parts Placement Layout

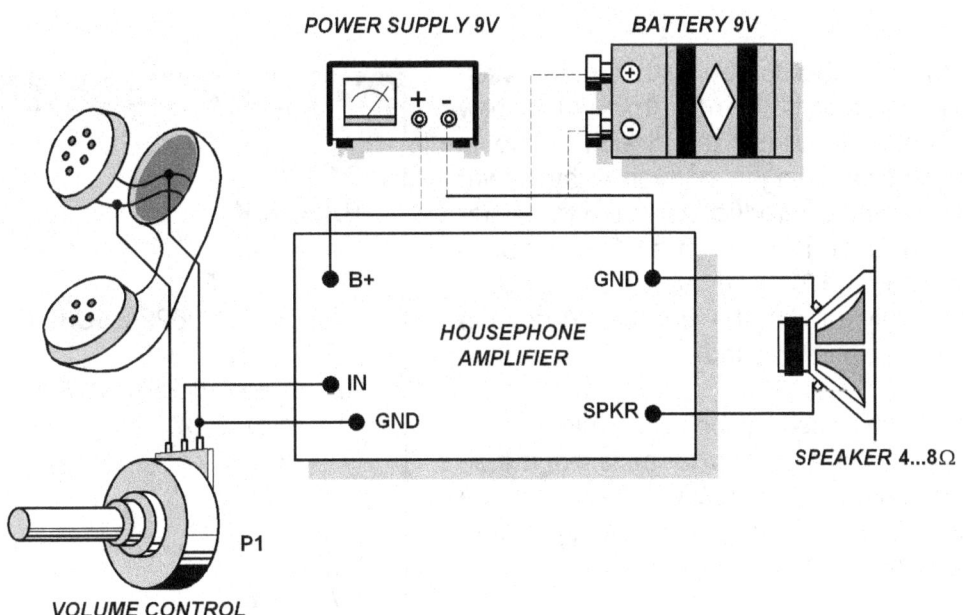

Figure 22.2 External Wirings

23 TOUCH DOORBELL

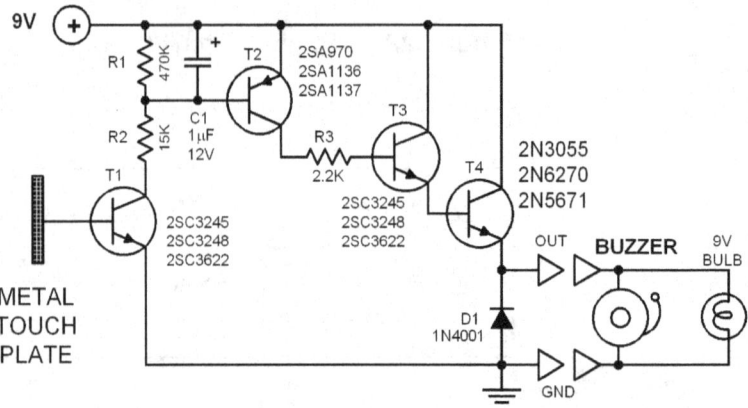

Diagram 23.0 Touch Doorbell

A highly reliable doorbell with "touch" switch can be easily contructed with few components. When the "touch" plate is touched by a finger, enough stray power line current is introduced into the transistor T1 so that it conducts. Succeeding transistors T2,T3,T4 also conduct and eventually trigger the doorbell (the bulb lights if one is installed).

The touch switch can be constructed from a copper plate, aluminum foil or similar metallic materials. The supply voltage can be between 6 and 12 volts. The standby current is around 6 mA.

Obviously, the circuit can also be applied as an alarm device for metallic objects.

Parts List
R1 = 470K
R2 = 22K
R3 = 2.2K
C1= 1µF/12V
T1,T3= 2SC3622/2SC3245/ 2SC3248
T2 = 2SA970/2SA1136/ 2SA2137
D1= 1N4001
Misc:
9V bulb
9V Buzzer or Gong

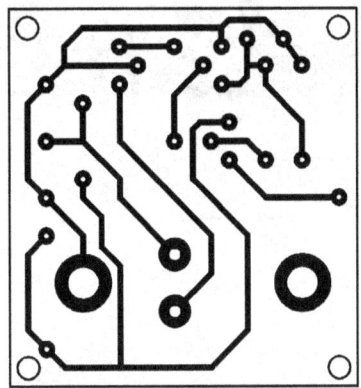

Figure 23.0
Printed Circuit Layout

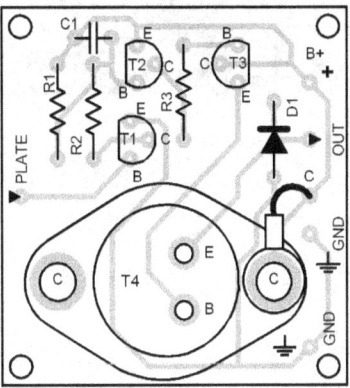

Figure 23.1
Parts Placement Layout

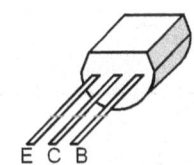

2SA970	2SC3622
2SA1136	2SC3245
2SA1137	2SC3248

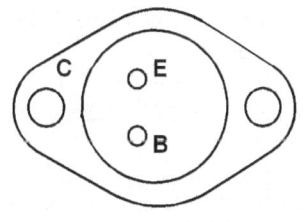

Bottom view

2N3055
2N6270
2N5671

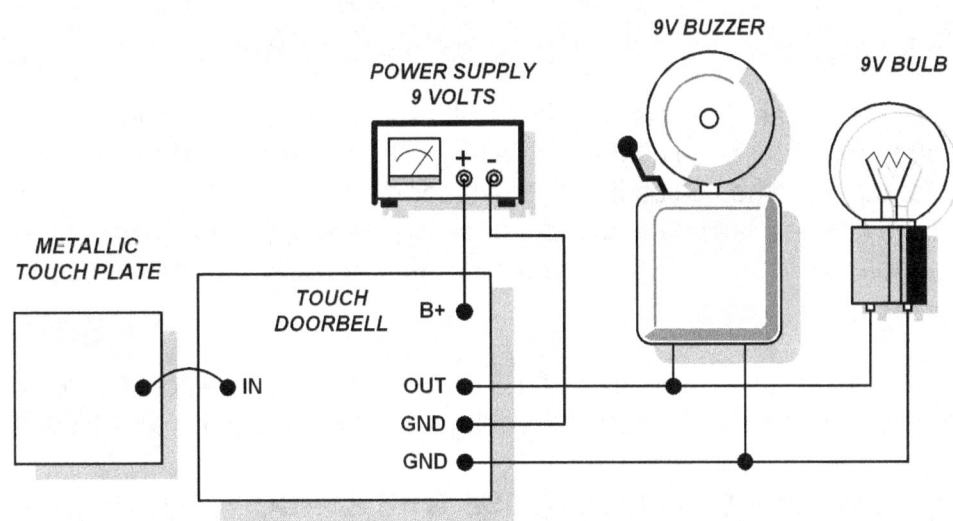

Figure 23.2 External Wiring

24   WATER LEVEL ALARM

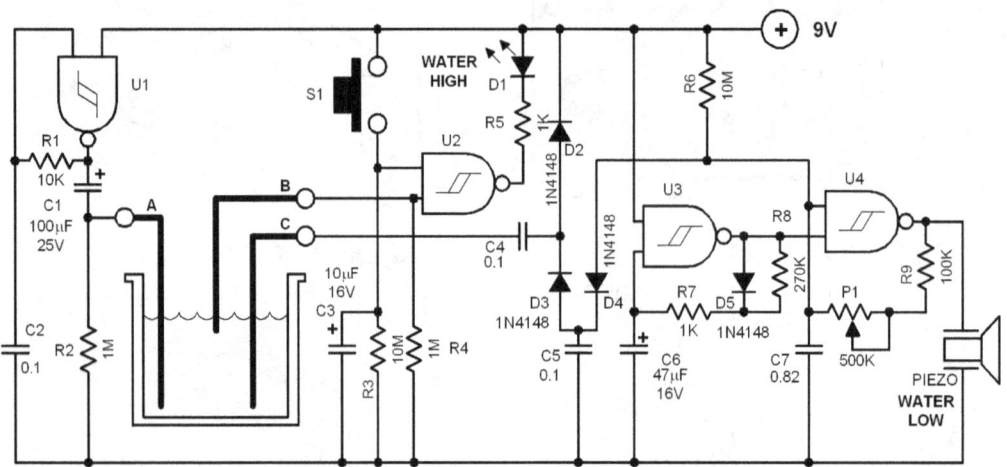

Diagram 24.0   Water Level Alarm

This circuit monitors the water level in a hydroculturepot. Once the water level is below electrode C, the circuit beeps periodically. If the pot is filled with water, the beeping stops immediately and when the water reaches electrode B, LED D1 lights signalling that the pot is already full with water. In order to conserve battery power, the LED will turn off automatically after 10 seconds.

Of course, you can use this circuit for other more important applications like monitoring the level of a liquid inside a sealed tank. The liquid must, however, be conductive. Some liquids like oil are poor conductors. One word of caution though, never use this circuit to monitor gasoline, alcohol or any other combustible/ inflammable liquids. The author takes no responsibility for any damages that may arise.

The beeper is a piezo-ceramic crystal. The electrodes can be constructed out of simple electrical wires. The supply voltage can be between 5 and 15 volts.

  **NEVER USE THE CIRCUIT TO MONITOR COMBUSTIBLE/INFLAMMABLE LIQUIDS LIKE GASOLINE, ALCOHOL, ETC.**

25 UNIVERSAL BEEPER

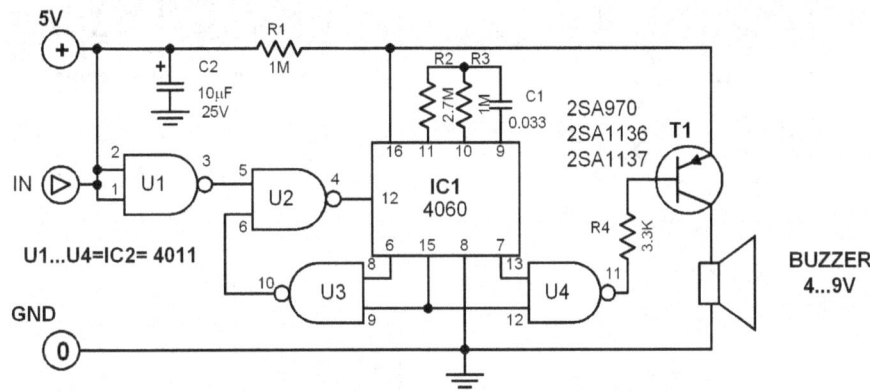

Diagram 25.0 Universal Beeper

This beep generator produces a periodic signal that has a universal application. It can be used as an acoustic signaller for almost any kind of monitoring circuits such as a doorbell, a limit sensor, an alarm clock, etc. The circuit begins to operate when the input logic drops to 0. After 30 seconds it will beep four times. When the input logic is not changed after 30 seconds it will beep again 4 times. This beeping cycle repeats continously until the input logic is raised back to 1. The heart of the circuit is a 14-stage binary counter with an internal oscillator. The frequency (f) is determined by C1 and R3. It can be found with:

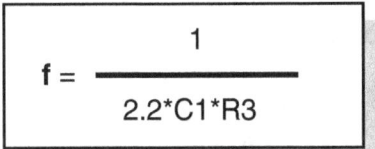

$$f = \frac{1}{2.2 * C1 * R3}$$

Parts List

R1 ,R3 = 1M
R2   = 2.7M
R4   = 3.3K
R7,R8= 8.2 ohms
C1 = 0.033/50V ceramic
C2 =   10µF/25V
T1 = 2SA970,2SA1136
IC1 =   4060
IC2 =   4011

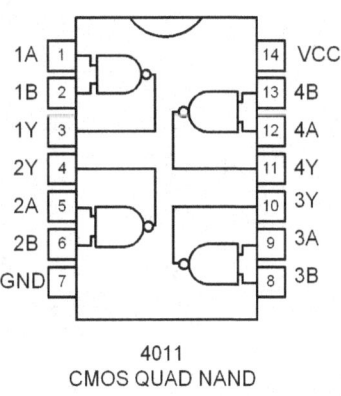

4011
CMOS QUAD NAND

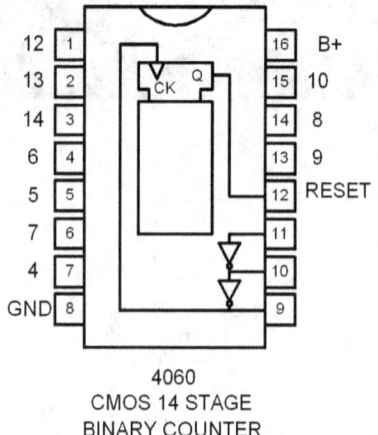

4060
CMOS 14 STAGE
BINARY COUNTER

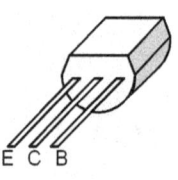

E C B

2SA970
2SA1136
2SA1137

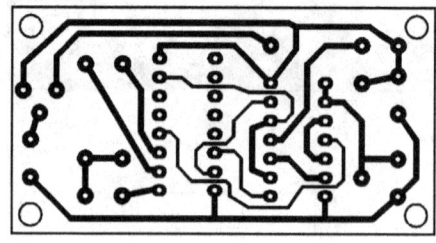

Figure 25.0   Printed Circuit Layout

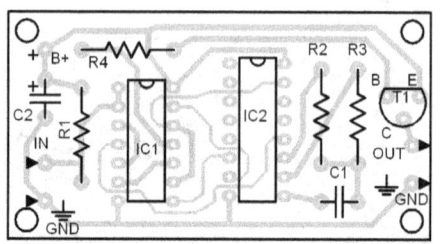

Figure 25.1   Parts Placement Layout

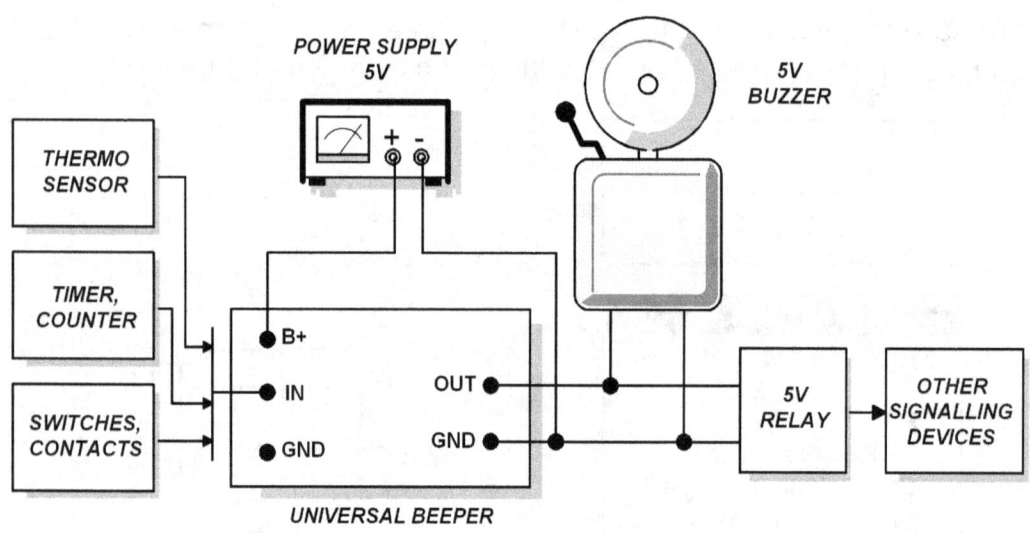

Figure 25.2   External Wirings

26 DOUBLE ALARM CIRCUIT

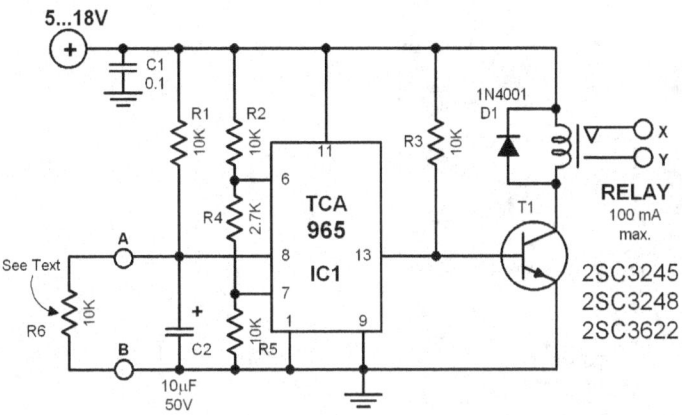

Diagram 26.0   Double Alarm Circuit

Alarm devices that use contacts as sensors activate when either a contact is broken or a contact is closed. When somebody opens a window or a door, a contact opens and the alarm activates. The opposite can also happen, a contact can close when a door or a window is forced open.

The alarm circuit featured here uses both principles to offer double security. Once the voltage at pin 8 becomes higher than at pin 7, the alarm activates. This happens when resistor R6 is disconnected either through the opening of a contact in series to the resistor or through the burglar's attempt to cut off the resistor sensor.

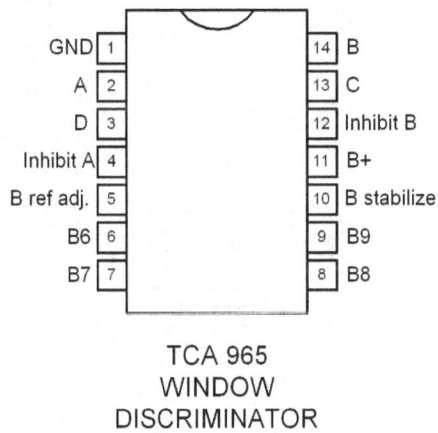

TCA 965
WINDOW
DISCRIMINATOR

The alarm also activates when resistor R6 is short circuited (contact parallel to resistor closes or when the burglar attempts to short circuit the resistor).

R6 can be connected either in parallel or series according to the type of alarm activation. Figure 26.2 shows some possible contact configurations. Since the wires used to interconnect all contact switches have also a resistance, the resistor R2 must be selected to give a total resistance of 10K.

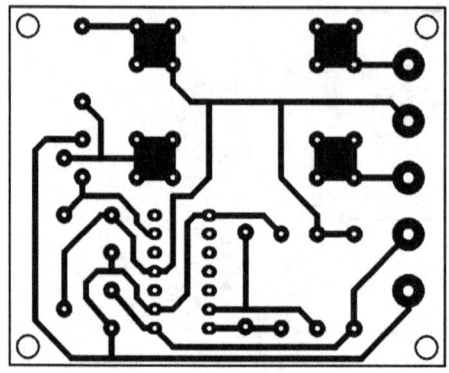

Figure 26.0 Printed Circuit Layout

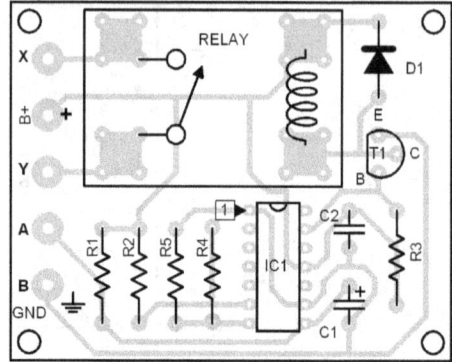

Figure 26.1 Parts Placement Layout

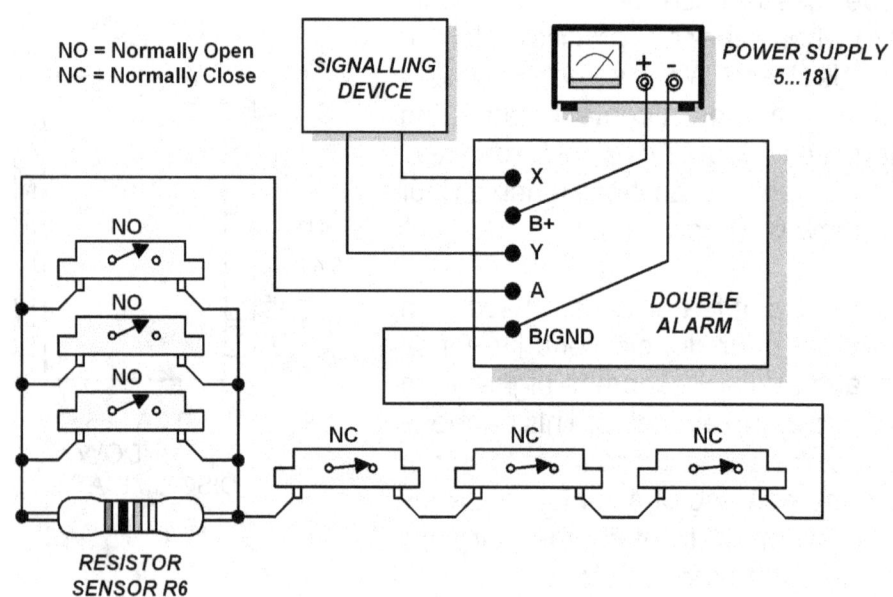

Figure 26.2 Sample External Wiring

27   3-DIGIT LOCK

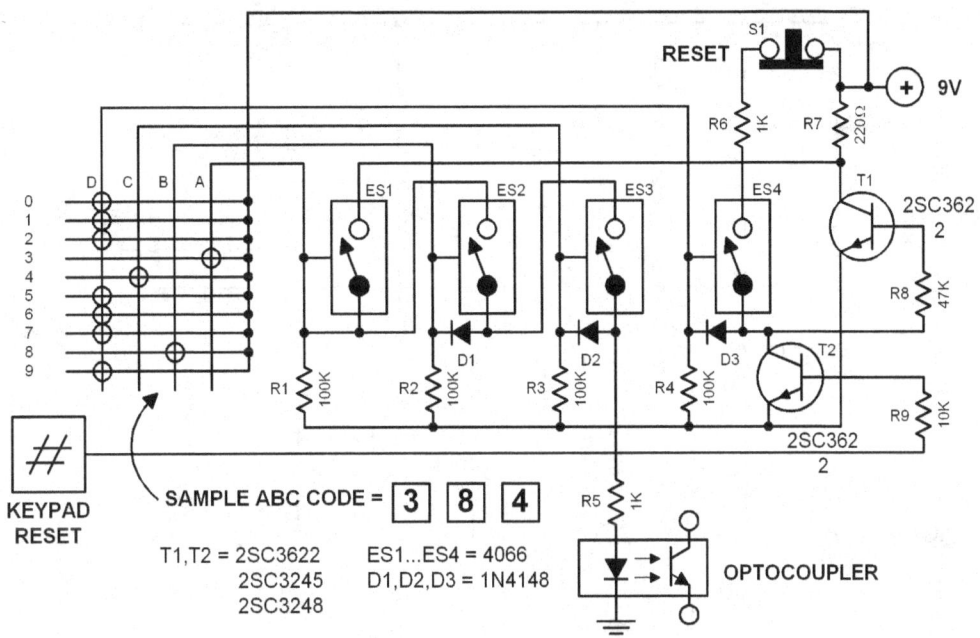

Diagram 27.0   3-Digit Lock

This electronic lock uses three series-wired analog switches for the key combination. These analog switches are connected to a programming matrix which is in turn controlled by the decimal keyboard. Let us take for example that wire A is connected to key 3, wire B to key 8 and wire C to key 4. Once key 3 is pressed, the analog switch ES1 closes and remains closed as long as current flows through R7. Then when key 8 is pressed, switch ES2 closes and remains closed unless ES1 opens. Finally key 4 is pressed and ES3 closes. The optocoupler activates and the phototransistor conducts. The phototransistor then activates any electromechanical device connected to It. Normally, the phototransistor drives a relay to increase its current capacity.

All other keys which are not connected to wires ABC are connected to wire D. Line D acts as a "jam" line. It disables the series-wired analog switches and acts as a security line. When somebody who doesn't know the correct combination presses a wrong key, line D activates and disables all analog switches. This makes the lock impossible to be opened by strangers. Once a wrong key is pressed, ES4 will close and T1 conducts thereby shorting the current line of ES1. The circuit is now jammed and cannot be opened even if the correct combination is entered later by chance. The circuit can only be reset by pressing S1 or the # key.

28   TELEPHONE RINGER

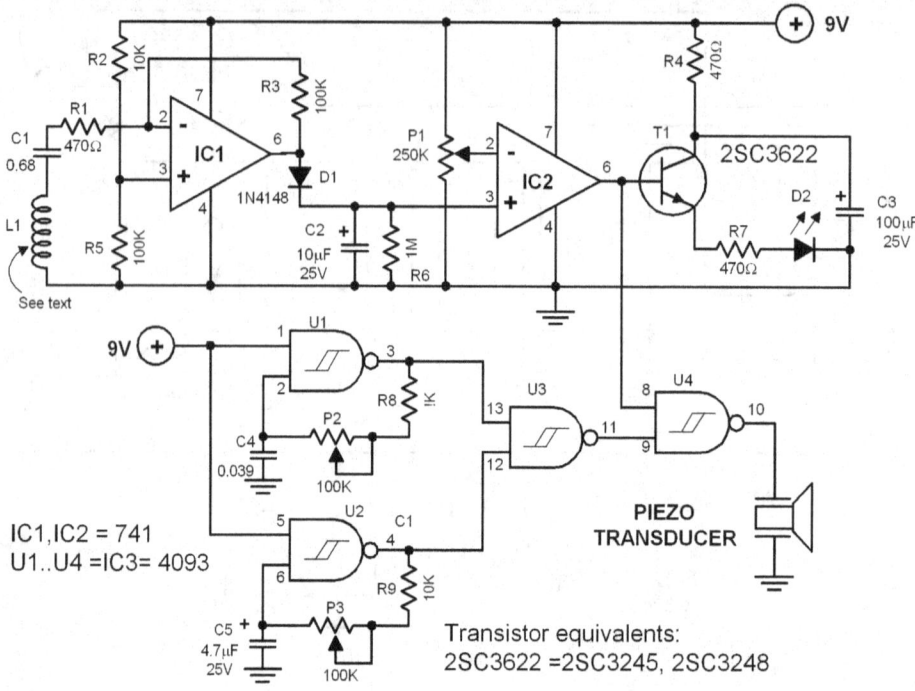

Diagram 28.0   Telephone Ringer

This circuit works as an auxiliary ringer providing a unique ring sound. It is not intended to replace the built-in ringer of a telephone. This is very helpful in a big office where several telephones are in use. In such a situation it is often difficult to determine at once which telephone rings. The circuit featured here provides a distinctive sound for each telephone set. The pitch and interval of the tone is adjustable.

The circuit uses an inductive coil L1 to pick up the ring signal from the telephone set. This signal is then amplified by IC1 and rectified by D1. This rectified signal is then fed to the comparator IC2. When the signal level is higher than the threshold level the comparator output rises to logic 1, and triggers the transistor T1 to conduction. The threshold level is adjustable through P1. Gate U4 is then switched on, and lets out the ringer tone generated by the astable multivibrator (U1 & U2) into the piezo-electric transducer.

The pitch of the tone is adjustable through P2. The tone interval is dependent on the setting of P3. Coil L1 is about 100 turns of fine magnet wire (size not critical) around a 10 cm x 10 cm square carton- see illustration. The coil must be placed under the telephone set (see illustration).

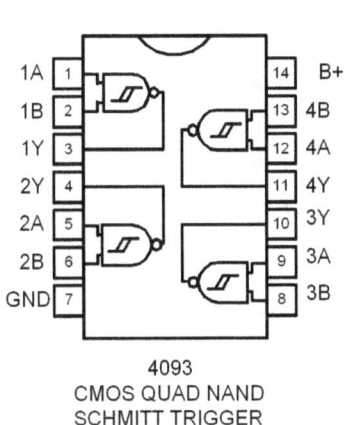

4093
CMOS QUAD NAND
SCHMITT TRIGGER

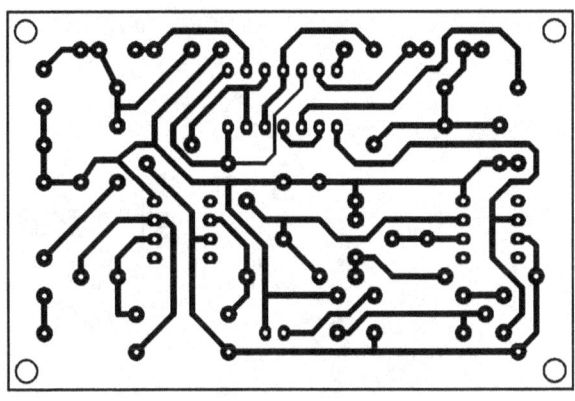

Figure 28.0   Printed Circuit Layout

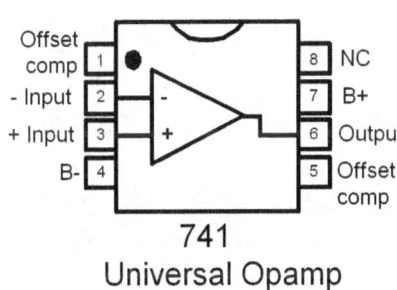

741
Universal Opamp

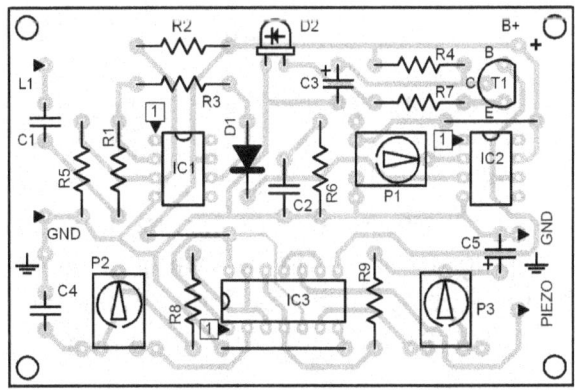

Figure 28.1   Parts Placement Layout

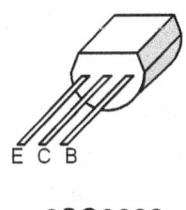

2SC3622
2SC3245
2SC3248

Pickup coil
(100 turns/10cm x 10cm

Coil under the
telephone set

29 CAR LAMP MONITOR

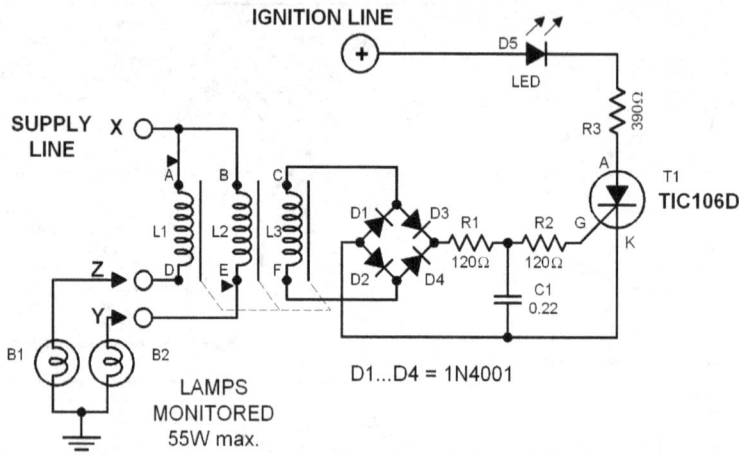

Diagram 29.0   Car Lamp Monitor

This circuit enables the driver to know if one of his rear lamps is not functioning properly without getting off and walking around the car. This is very helpful during driving. The circuit can monitor two lamps at a time. It monitors the lamps by sensing the magnetic field induced by the current flowing into the wires. The two windings L1 and L2 in the sensor coil are wound counter to each other. When both lamps being monitored function properly the magnetic field induced by the coils cancel each other so that no current is induced in the third coil L3.

However once one lamp burns out, the magnetic field changes and induces current in the third coil. This current triggers thyristor T1 into conduction and lights the LED signalling that one of the lamps is burned out. The LED remains lighted until the supply current or the ignition key is turned off.

The size of the magnet wire for the sensor coils must be at least 0.7 mm diameter. The ferrite ring is an ordinary core commonly found in lamp dimmers.

Parts List
R1,R2 = 220Ω
R3   = 390Ω
C1,C2= 0.22/50V ceramic
D1,D2,D3,D4 = 1N4001
D5 = LED
T1 = TIC106D (>1A/>50V)

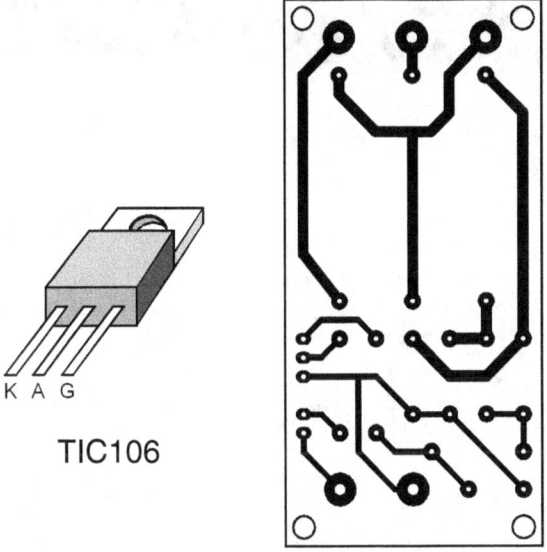

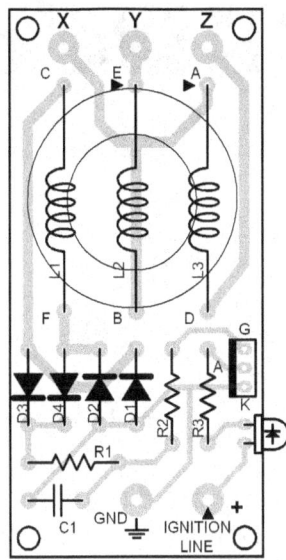

TIC106

Figure 29.0
Printed Circuit Layout

Figure 29.1
Parts Placement Layout

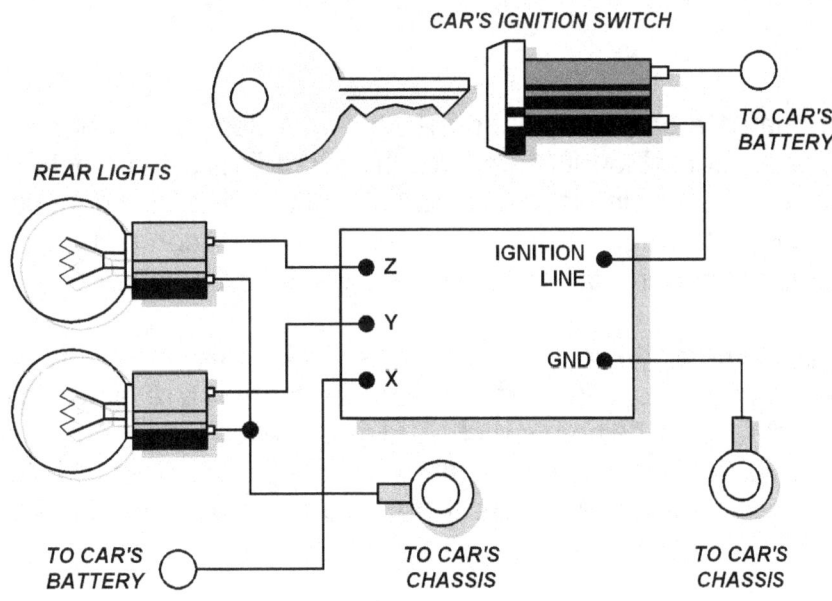

Figure 29.2   External Wiring

30   CAR RADIO ALARM

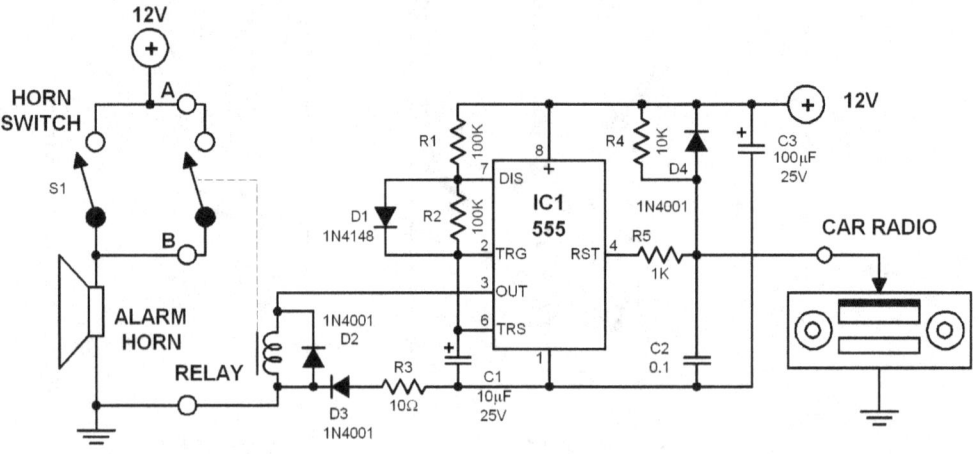

Diagram 30.0   Car Radio Alarm

This alarm guards a car radio round-the-clock. Its design is very simple. The circuit causes the car to blow its horn once the car radio is disconnected from the car chassis. The trick is simple, a wire from the alarm circuit is connected to the metallic housing of the car radio. Once this wire is disconnected, the alarm activates, and the car produces its own version of a horn concert.

The circuit is a simple 555 timer wired as an astable multivibrator. Its frequency is determined by C1. The output of the IC directly controls a relay which is in turn connected in parallel to the horn switch. The reset switch of the IC is used as the sensor point. It is connected to the metal housing of the car radio. This technique makes it impossible for the thief to steal the radio away without taking the alarm circuit with him.

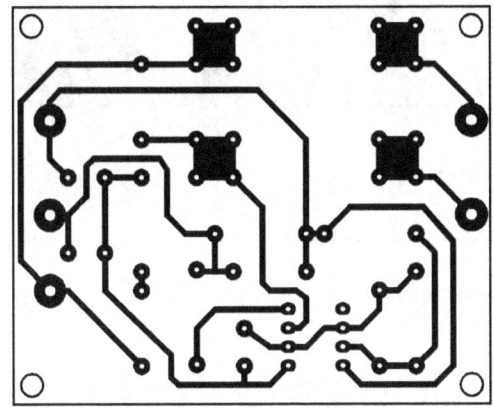

Figure 30.0 Printed Circuit Layout

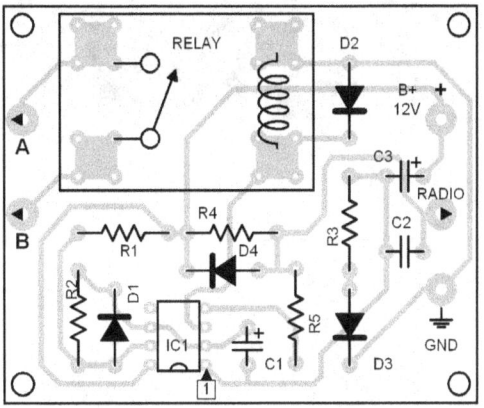

Figure 30.1 Parts Placement Layout

Parts List
R1,R4= 4.7K
R2,R3= 68K
R5,R6= 5.6K
R7,R8= 8.2 ohms
C1,C2= 10µF/10V
C3=   2.2µF/25V
C4,C5= 22µF
C6,C7= 220µF/10V
D1,D2= 1N4148
IC1=   TDA 1522

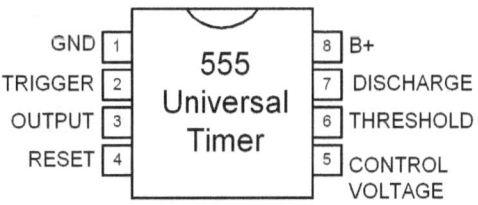

31   HIGH HEAT MONITOR

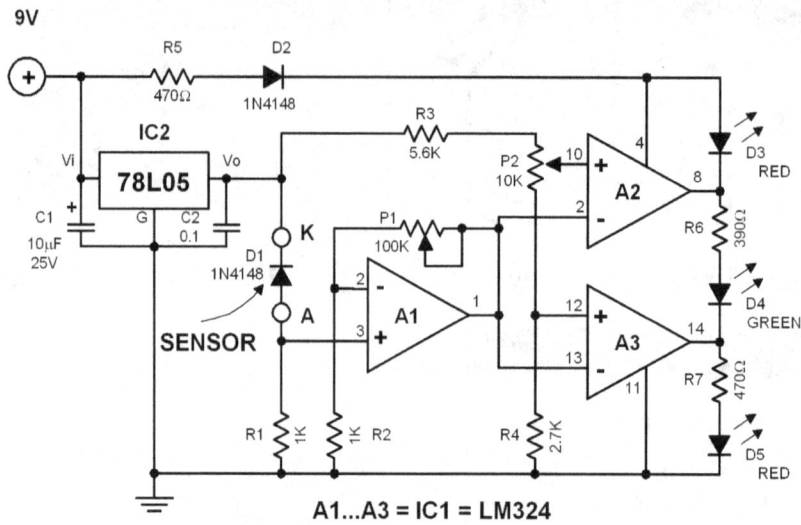

Diagram 31.0   High Heat Monitor

This circuit is designed to monitor the temperature of cooking oil for foods which highly depend on the temperature constancy to obtain the right taste like french fries for example.

The temperature sensor used is a common 1N4148 diode. The reverse current of this diode changes logarithmically according to the temperature. At temperatures around 200°C, this current is strong enough to be measured. The reverse current is converted to a voltage level and amplified by A1. This amplified voltage is fed to the negative inputs of A2 and A3. At the same time, a reference voltage is fed to the plus inputs of A2 and A3.

IC2 (low power regulator) stabilizes this reference voltage so that the function of the circuit will not be affected by the condition of the battery.

To calibrate the circuit, you need two heat sources, one 175°C and a 185°C. Of course, you also need a high temperature thermometer (up to 200°C). Initially, dip the diode sensor into the 175°C hot water (or oil), wait for a few seconds and adjust P1 slowly until D4 turns off and D5 turns on. Finally, dip the diode sensor into the 185°C liquid and slowly adjust P2 until D5 turns off and D3 turns on. Use high temperature wires to connect the diode to the circuit. Use crimp connectors to mechanically fix the diode to these wires. Never solder it, otherwise, you will get your fries spiced up with molten solder. Solder melts easily in these temperature levels.

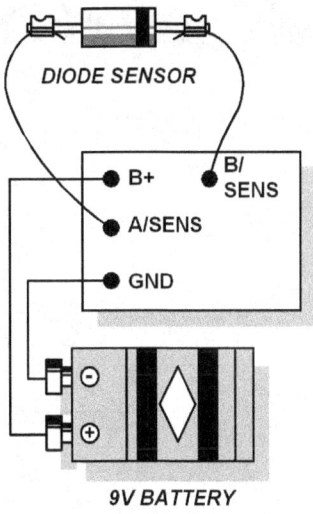

Figure 31.2 External Wiring

Use crimp connectors to fix the diodes to the wires. Never solder it!

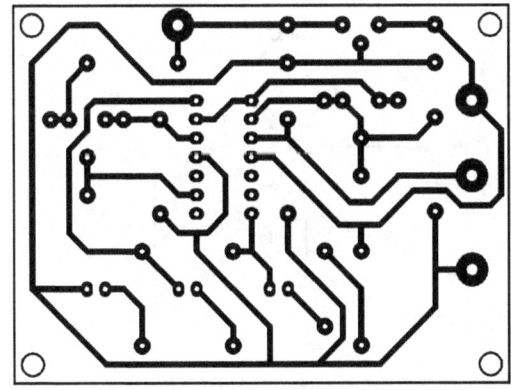

Figure 31.0 Printed Circuit Layout

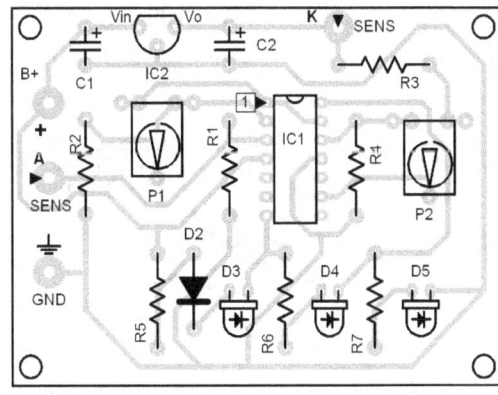

Figure 31.1 Parts Placement

32   MULTI-SOUND SIREN

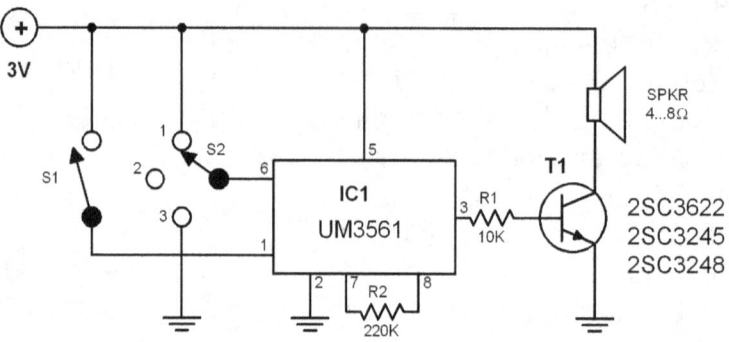

Diagram 32.0   Multi-Sound Siren

This siren can generate four types of siren sounds and is composed of only two active components. The siren sound can be selected through the combination of two switches (see Table 32.0). This circuit is universal and can be applied as a signalling device for alarms, monitors, doorbells, pagers, etc.

S2	S1	SOUND
2	OFF	POLICE
1	OFF	FIRE
3	OFF	AMBULANCE
1	ON	STACCATO

Table 32.0

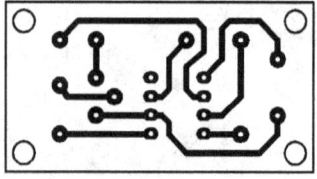

Figure 32.0   Printed Circuit Layout

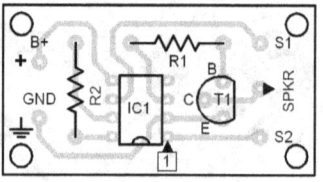

Figure 32.1   Parts Placement

33 CAR ALARM

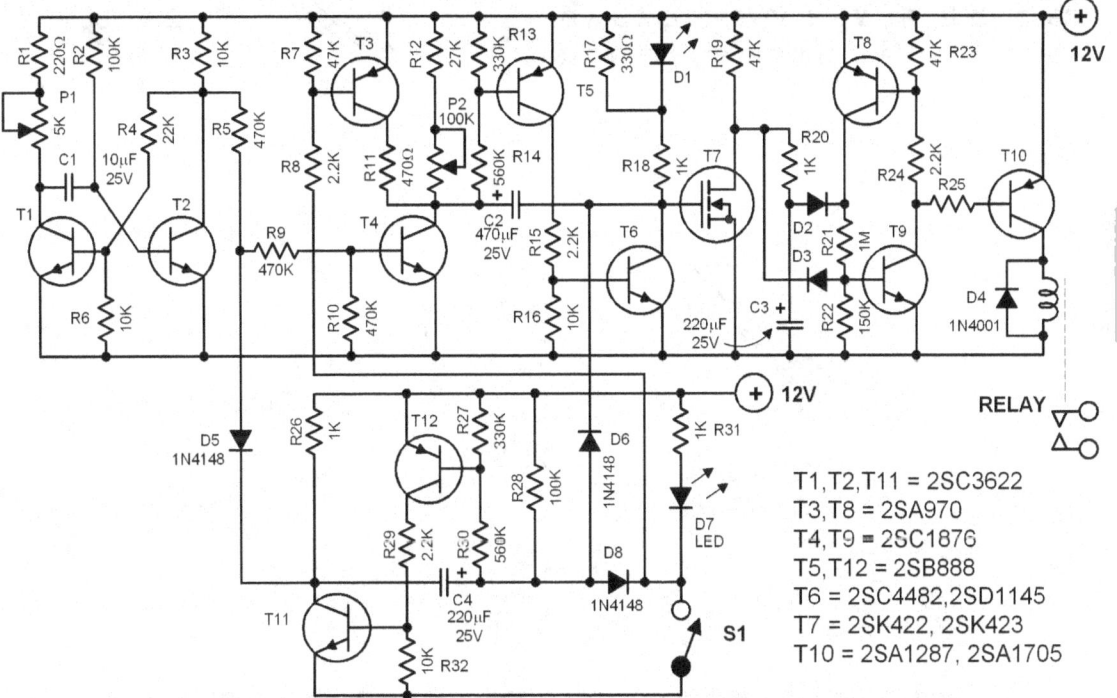

Diagram 33.0 Car Alarm

T1,T2,T11 = 2SC3622
T3,T8 = 2SA970
T4,T9 = 2SC1876
T5,T12 = 2SB888
T6 = 2SC4482,2SD1145
T7 = 2SK422, 2SK423
T10 = 2SA1287, 2SA1705

This circuit monitors the supply line of the car. It reacts to any kind of short voltage changes which is normally generated by switching on a current load like the switching on of a lamp when the door is opened.

This alarm circuit offers the advantage of simplicity in installation. You don't need to install alarm switches around your car. Simply connect the circuit to the car's battery and chassis, and your car is guarded right away. No need to change or add any wirings.

Once activated, the alarm closes a relay that is connected in parallel to the horn switch. The alarm is put into operation by opening S1. It is delayed by 25 seconds to allow the driver to get out of the car before it completely becomes "active ".

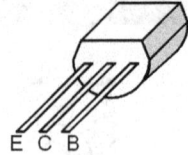

2SA970
2SA1287
2SB888
2SC3622
2SD1145

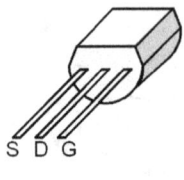

2SK422
2SK423

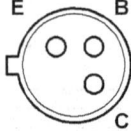

2SC1876

Bottom view

2SA1705
2SC4482

Parts List:

R1 = 220Ω
R2,R28 = 100K
R3,R6,R16,R32 = 10K
R4 = 22K
R5,R9,R10 = 470K
R7,R19,R23 = 47K
R8,R15,R24,R25,R29 = 2.2K
R11 = 470Ω
R12 = 27K
R13,R27 = 330K
R14,R30 = 560K
R17 = 330Ω
R18,R20,R26,R31 = 1K
R21 = 1M
R22 = 150K

C1 = 10µF/25V
C2 = 470µF/25V
C3,C4 = 220µF/25V
D1,D7 = LED
D2,D3,D3,D4,D6,D8= 1N4148

T1,T2,T11 = 2SC3622
T3,T8 = 2SA970
T4,T9 = 2SC1876
T5,T12 = 2SB888
T6 = 2SC4482, 2SD1145
T7 = 2SK422, 2SK423
T10 = 2SA1287, 2SA1705

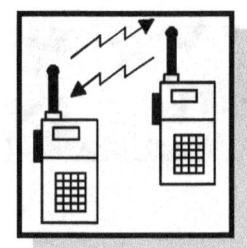

RADIO FREQUENCY

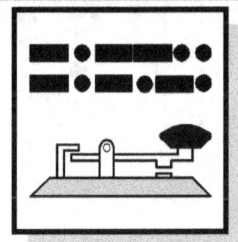

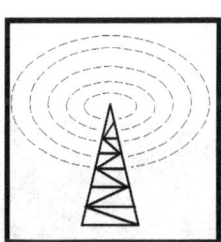

34 MINI AM RADIO

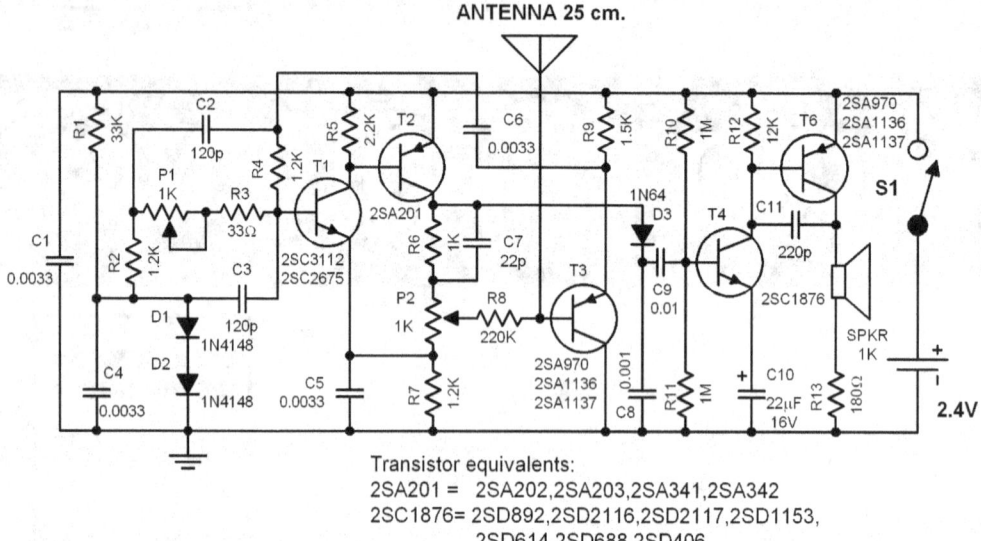

ANTENNA 25 cm.

Transistor equivalents:
2SA201 = 2SA202,2SA203,2SA341,2SA342
2SC1876= 2SD892,2SD2116,2SD2117,2SD1153,
2SD614,2SD688,2SD406

Diagram 34.0 Mini AM Radio

This radio circuit uses a filter for the frequency selection. The design makes the circuit very compact since it does not need a variable capacitor nor ferrite rod. The circuit can be constructed in matchbox size. Since there is no variable capacitor, the frequency must be set through P1. The antenna must be 25 cm long.

This radio works at its best when it is held in hand or placed on top of a metallic surface. Potentiometer P2 sets the maximum sensitivity of the radio. The current consumption is around 2 mA.

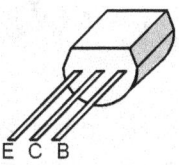

E C B

2SA970   2SA1136
2SA1137   2SC3112
2SC2675   2SD1153
2SD892

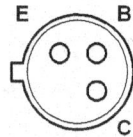

E   B

C

2SA201   2SD614
2SA202   2SD406
2SA203   2SD688
2SC1876

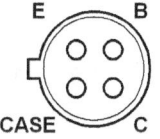

E   B

CASE   C

2SA341
2SA342

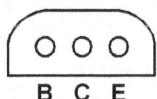

B C E

Bottom view

2SC2116
2SC2117

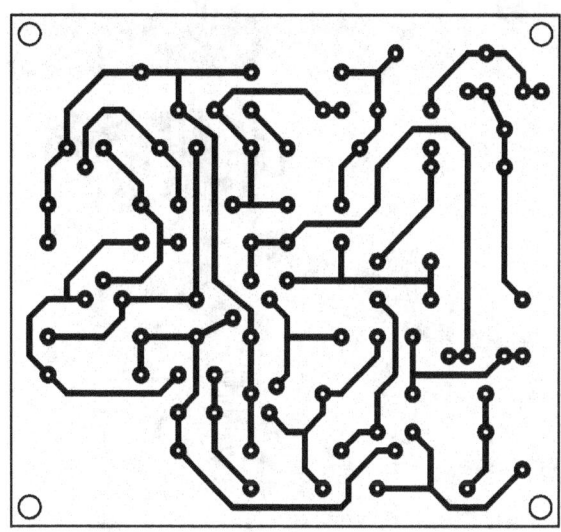

Figure 34.0   Printed Circuit Layout

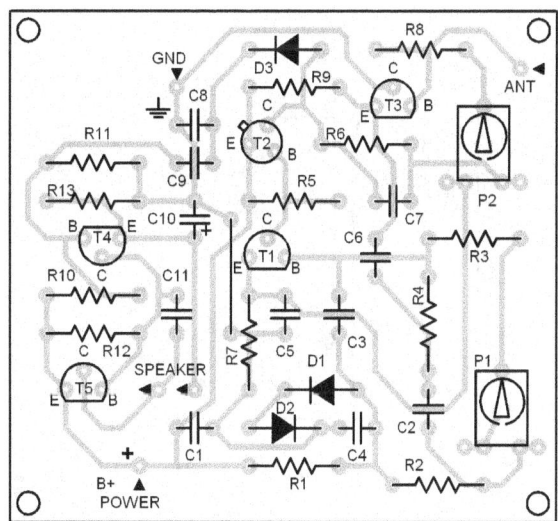

Figure 34.1   Parts Placement Layout

35 FM-CB RECEIVER

FM-CB is becoming more and more popular because of its better reception quality compared to the antique AM mode. FM reception is cleaner, stronger and has a "capture" effect.

This receiver is a one-way superheterodyne receiver with an IF of 455 kHz. The receiver filter is a 27MF type so that the receiver does not need to be adjusted or calibrated. The demodulation of the FM signal is done with a ceramic filter in a discriminator circuit.

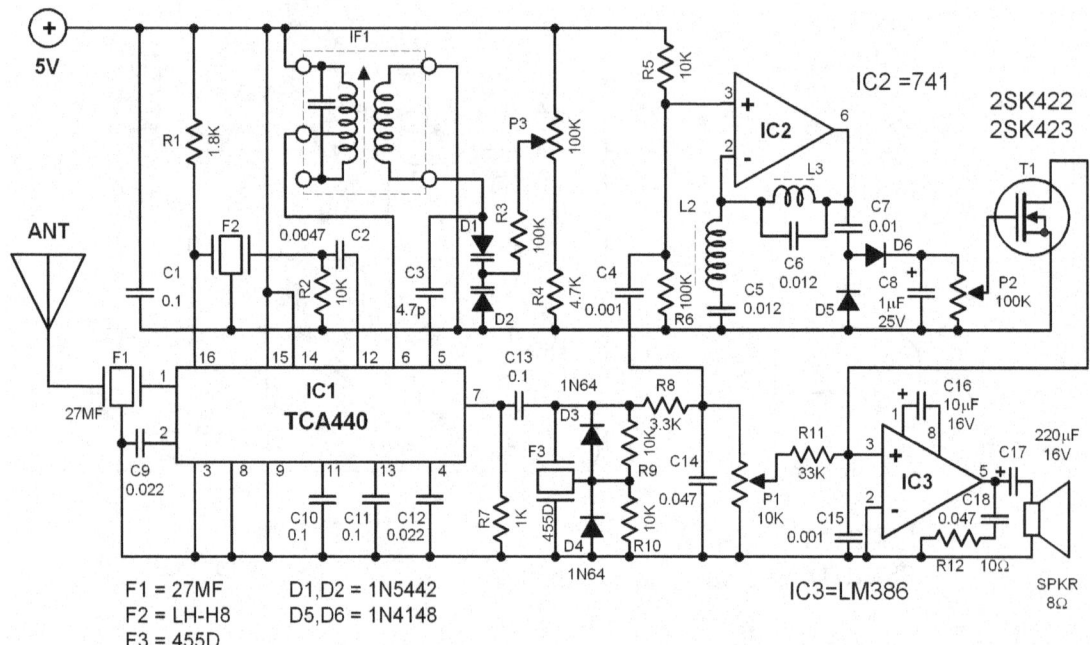

F1 = 27MF   D1,D2 = 1N5442
F2 = LH-H8   D5,D6 = 1N4148
F3 = 455D

Diagram 35.0   FM-CB Radio

36 RTTY CONVERTER

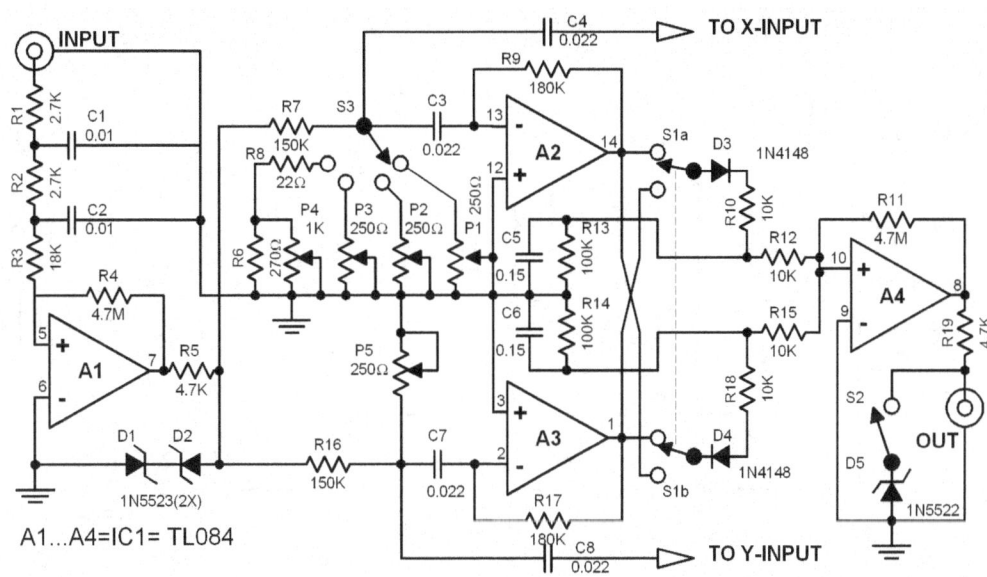

Diagram 36.0   RTTY Converter

RTTY means Radio TeleTYpe. It is one form of data transmission, and is done with different codes. An important code is the Baudot code, that is why this circuit is designed to receive RTTY signals in baudot form.

The circuit consists of one TL084 IC and few other components. The IC integrates four opamps that are enough to build the filter and limiter stages of this circuit. The circuit also generates control signals that can be to fed to the X and Y inputs of an oscilloscope that is used in aligning the filter stages. The speeds of the converter are 45 baud, 50 baud, 75 baud, and 110 baud.

The mark filter has a single center fequency of 1275 Hz, and can be fined tuned through trimmer P5. The space filter has a variable center frequency, and is selectable from 1445 Hz, 1700 Hz and 2125 Hz. These frequencies are fined tuned through P1, P2 and P3 respectively. The potentiometer P4 is used for shifting up the center frequency from 170 Hz to 1000 Hz. In using an oscilloscope to fine tune the filter stages, set the mark and space center frequencies until two ellipsoids form a cross in the center of the oscilloscope's screen.

37 TUNABLE ACTIVE ANTENNA

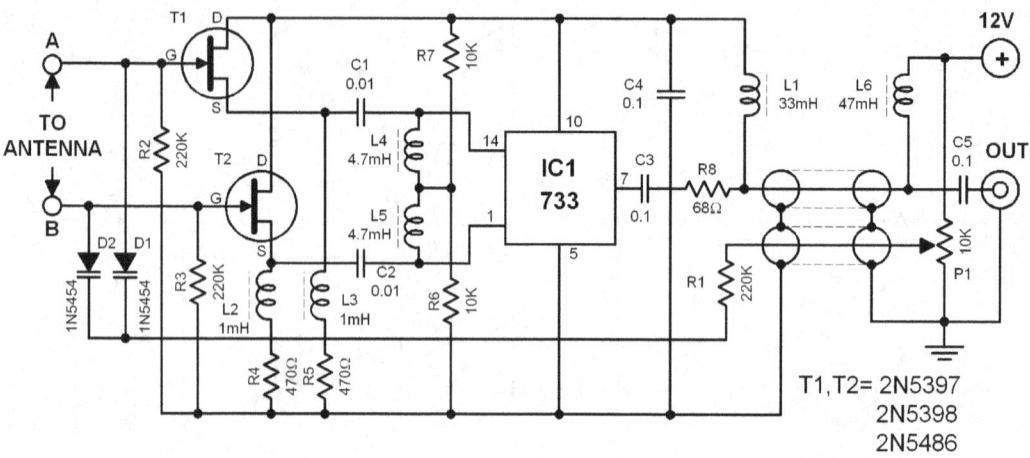

Diagram 37.0 Tunable Active Antenna

This circuit is used for shortwave receivers. It can be used for a loop or a dipole antenna. The capability to switch over from one frequency band to another is absent in this design to make the circuit easy to construct. It, however, offers a different advantage: If you are using a loop antenna, you don't need to shield it because the input of the circuit is symmetrical. The circumference of the loop antenna must not be longer than 10% of its wavelength.

If you use a dipole antenna, the total length of the dipole can be less than the wavelength or equal to 10% of the wavelength.

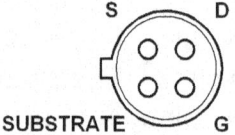

2N5397
2N5398

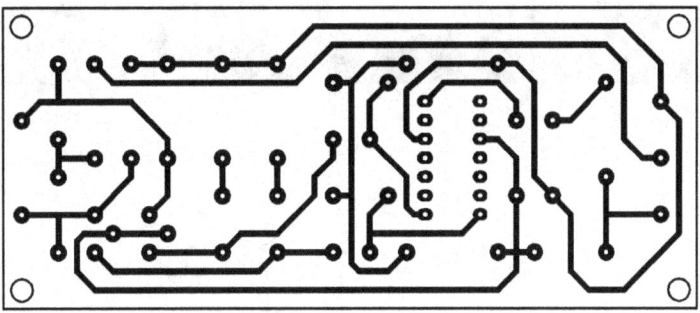

Figure 37.0   Printed Circuit Layout

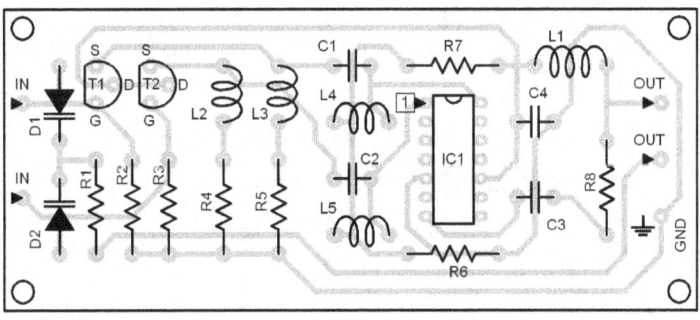

Figure 37.1   Parts Placement Layout

Parts List:		
Resistors:	Diodes:	
R1,R2,R3 = 220K	D1,D2= 1N5454	
R4,R5 = 470Ω		
R6,R7 = 10K	T1,T2 = 2N5486	
R8 = 68Ω		
P1 = 10K	Coils:	
	L1 = 33mH	
Capacitors:	L2,L3 = 1mH	
C1,C2= 0.01/50V	L4,L5 = 4.7mH	
C3,C4,C5= 0.1/50V	L6 = 47mH	

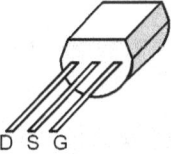

D S G

2N5486

38 VIDEO AMPLIFIER

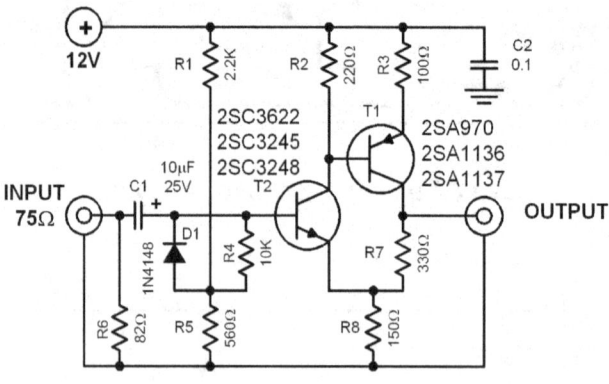

Diagram 38.0   Video Amplifier

This circuit amplifies a video signal to enable it to drive the video input of a portable B/W TV. This circuit was originally designed for television sets which are used as computers monitors. The amplifier has a video bandwidth of 10 MHz.

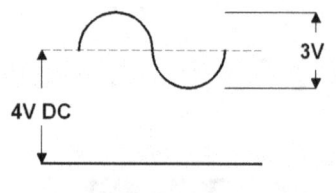

Output Signal Level

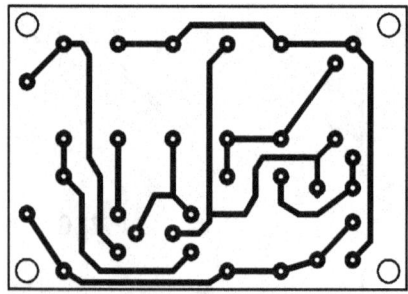

Figure 38.0   Printed Circuit Layout

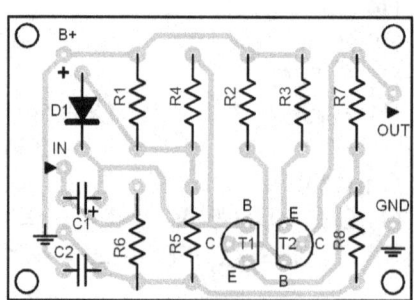

Figure 38.1   Parts Placement

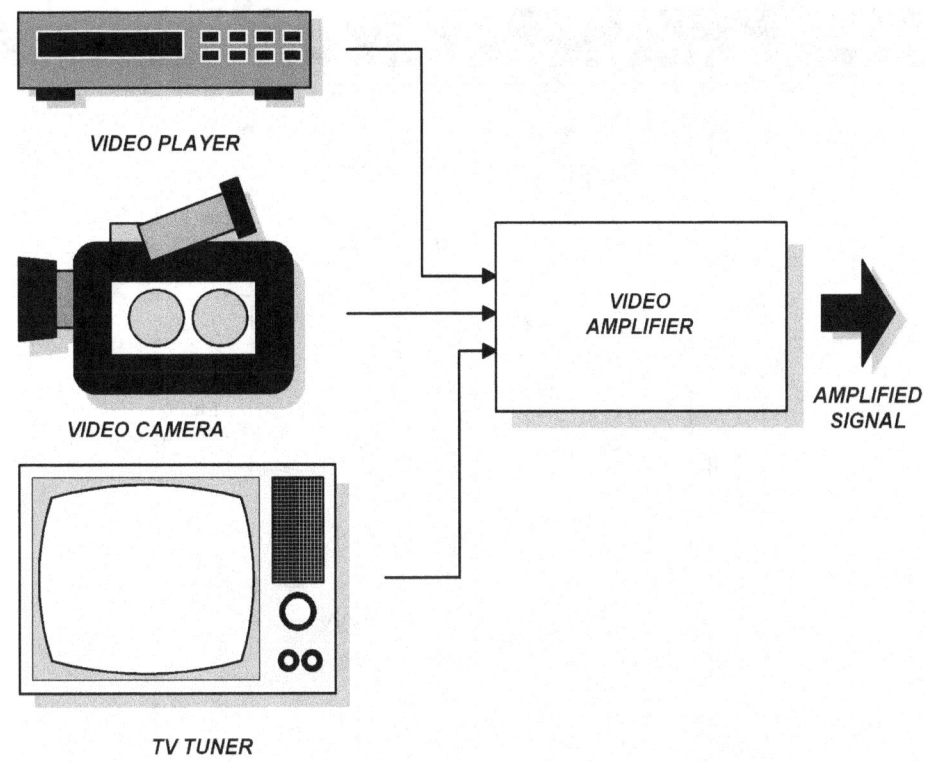

Figure 38.2 Sample Applications

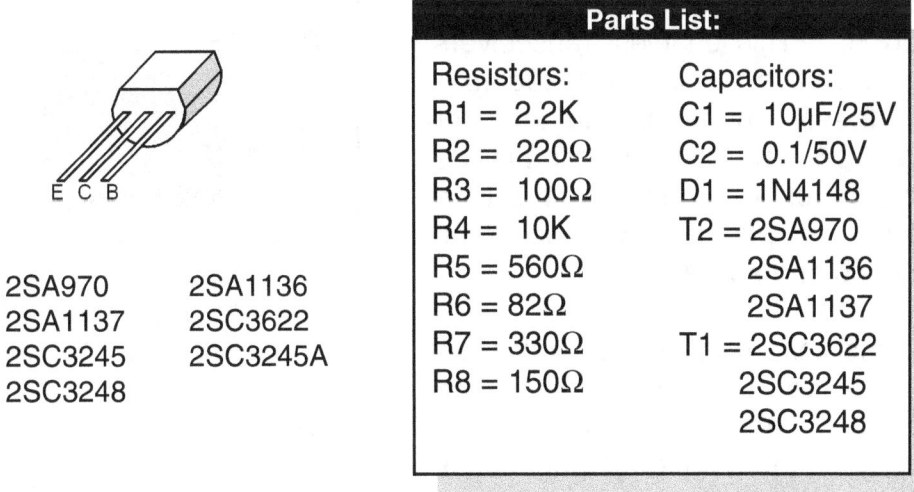

2SA970   2SA1136
2SA1137   2SC3622
2SC3245   2SC3245A
2SC3248

Parts List:	
Resistors:	Capacitors:
R1 = 2.2K	C1 = 10µF/25V
R2 = 220Ω	C2 = 0.1/50V
R3 = 100Ω	D1 = 1N4148
R4 = 10K	T2 = 2SA970
R5 = 560Ω	2SA1136
R6 = 82Ω	2SA1137
R7 = 330Ω	T1 = 2SC3622
R8 = 150Ω	2SC3245
	2SC3248

39 REPEATER ACCESS ENCODER

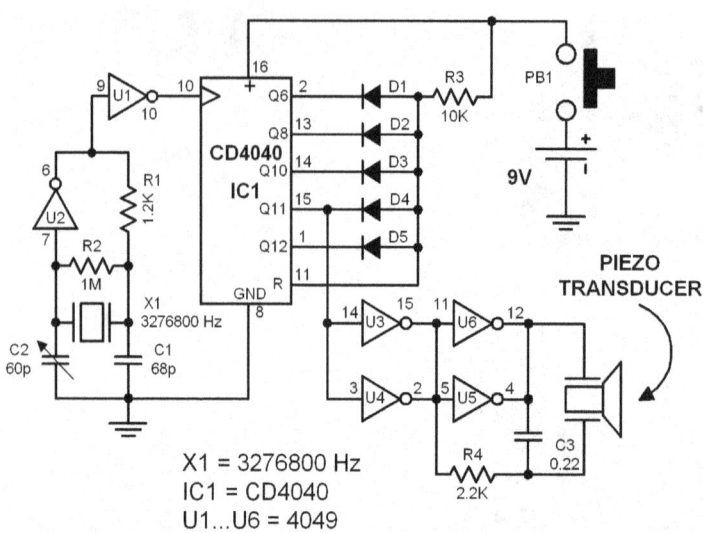

X1 = 3276800 Hz
IC1 = CD4040
U1...U6 = 4049

Diagram 39.0   Repeater Access Encoder

Some FM amateur repeaters can only be accessed by using a tone code of 1750 Hz. The circuit featured here generates a tone of 1750 Hz. This enables transceivers which do not have built-in tone encoders to access these repeaters. The encoder circuit doesn't need to be installed inside the transceiver. For simplicity, the tone encoder is just placed very close to the microphone and activated.

The correct tone frequency is generated from a divider IC which divides the frequency of the crystal. The desired frequency is then buffered and finally delivered into the piezo-electric crystal. The current consumption is around 5 mA.

Parts List:
R1 = 1.2K
R2 = 1M
R3 = 10K
R4 = 2.2K
C1 = 68p/50V
C2 =   60p trimmer capacitor
C4,C5= 22µF
C3 = 0.22/50V
D1,D2,D3,D4,D5 = 1N4148
IC1 =   4040
IC2 =   4049

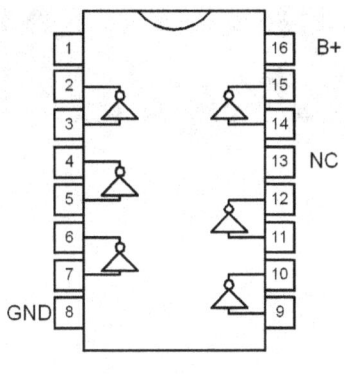

4049 (CMOS)
6 INVERTING BUFFERS

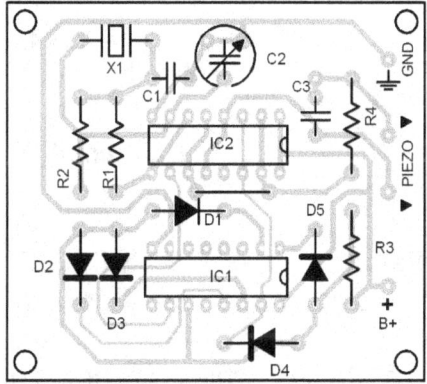

Figure 39.0   Printed Circuit Layout

4040 (CMOS)
ASYNCHRONOUS
12 STAGE BINARY
COUNTER

Figure 39.1   Parts Placement

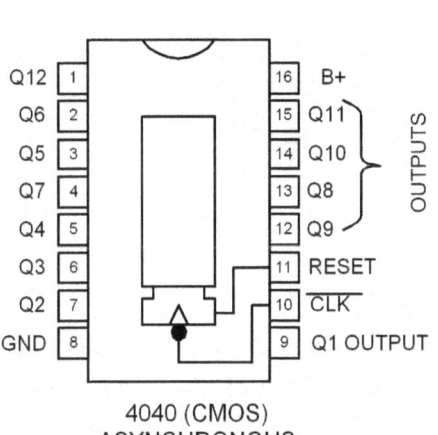

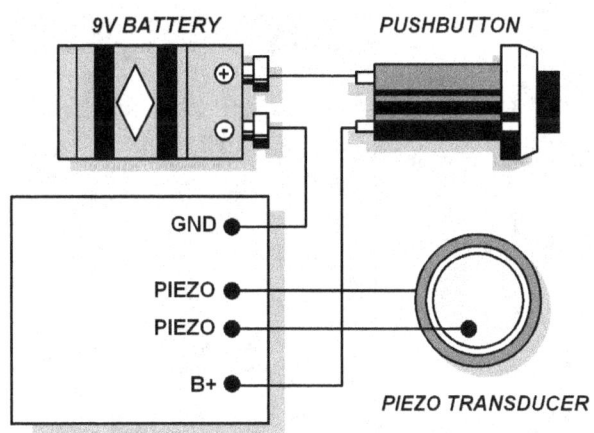

Figure 39.2   External Wiring

40 SATELLITE INTERFACE

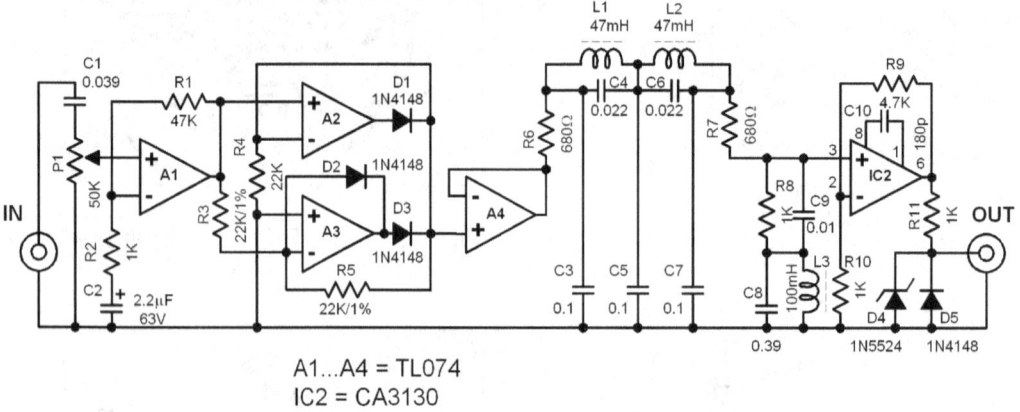

A1...A4 = TL074
IC2 = CA3130

Diagram 40.0 Satellite Interface

This interface circuit detects the AM signal of the weather satellite information transmitted in around 138 MHz. The actual weather data amplitude modulates an auxillary carrier of 2400 Hz. This auxillary carrier then FM modulates the main carrier of 138 MHz which is beamed down to earth. To decode the weather data, one needs a computer and an appropriate program.

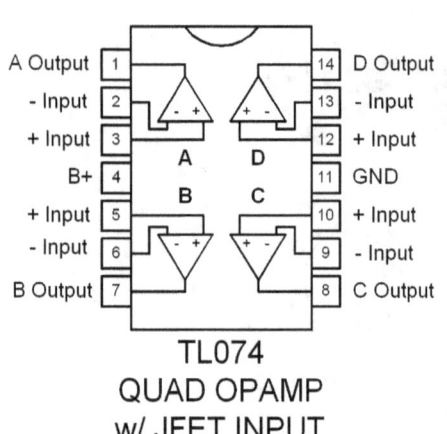

TL074
QUAD OPAMP
w/ JFET INPUT

Parts List:	
R1 = 47K	C6 = 0.018/50V
R2,R89,R10,R11= 1K	C9 = 0.01/50V
R3,R4,R5 = 22K	C10 = 180p/50V
R6,R7 = 680Ω	
R9 = 4.7K	IC1 = TL074
C1,C8= 0.039/50V	IC2 = CA3130
C2= 2.2µF/63V	
C3,C5,C7 = 0.1/50V	D1,D2,D3,D5 = 1N4148
C4 = 0.022/50V	D4 = 1N5524

HOBBY & GAMES

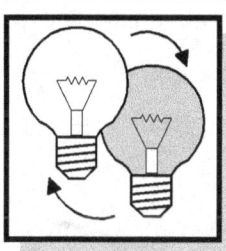

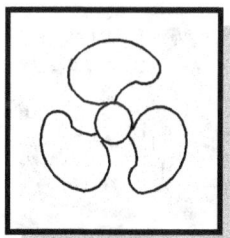

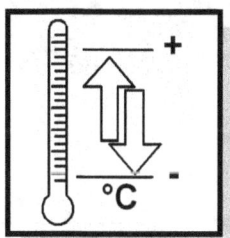

41 MODEL BRAKE LIGHT

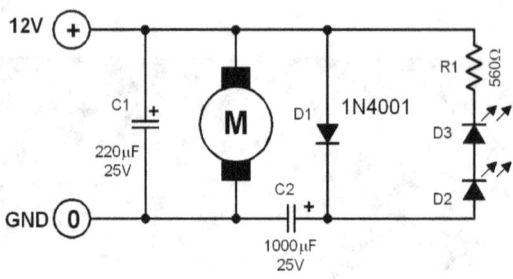

Diagram 41.0   Model Brake Light

This circuit creates a realistic brake light effect for model cars(or train). It works this way: when the model car is running, the batteries supply current to the motor and capacitors C1 and C2. Once the motor is turned off (stopping the model car), capacitor C1 discharges through the motor. At the same time capacitor C2 discharges through the two LEDs and the LEDs light up until the capacitor is fully discharged. R is 560 ohms when the supply voltage is 12 volts.

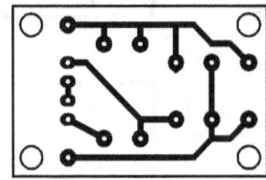

Figure 41.0   Printed Circuit Layout

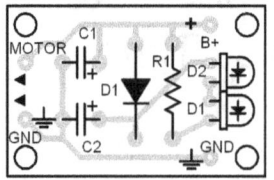

Figure 41.1   Parts Placement

42 LAP COUNTER

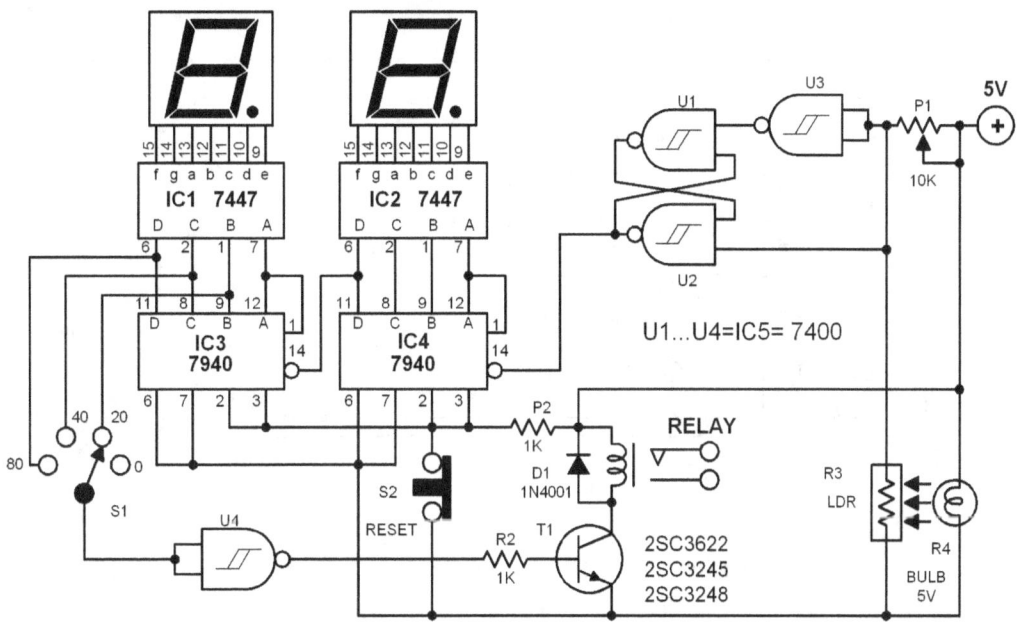

Diagram 42.0 Lap Counter

This circuit is used to count the laps in a model car racing. It uses a light switch in place of a troublesome mechanical switch to count the laps. The light switch is composed of a small bulb and an LDR. These two components are installed on the racing track in such a way that the light beam will be interrupted everytime a model car passes by. Definite lap counts like 20, 40 or 80 can be selected in advance through S1. Once the maximum lap count is reached, the corresponding outputs B, C, and D give out a logic 1 which is inverted by U4 and switches off T1.

Parts List:

R1,R2 = 1K
R3 = LDR (ligh dependent
resistor)
P1 = 10K
IC1,IC2 = 7447
IC3,IC4 = 7490
IC5 = 7400
T1 = 2SC3622/2SC3245
D1 = 1N4001
5V Bulb

The relay opens as a result and breaks the power line of the race track thereby stopping the race. The LDR must be shielded by a short piece of tubing so that stray light will not reach it. The sensitivity of the LDR can be adjusted through P1.

43 PHOTO EXPOSURE TIMER

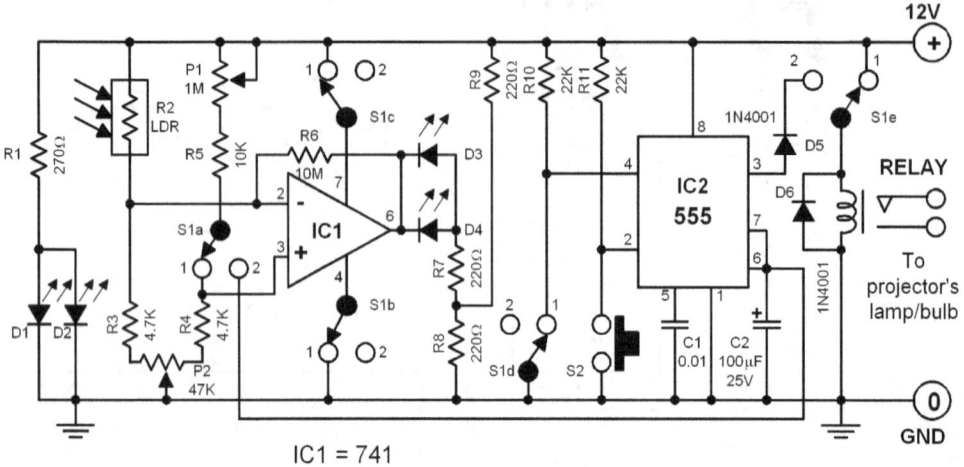

Diagram 43.0   Photo Exposure Timer

This circuit has a double function - it works as light intensity meter and as exposure timer. An LDR samples the intensity of the light coming from the enlarger and converts it to electrical current. This current is fed to the bridge circuit which is in turn connected to the inputs of the IC1. The balance of the bridge circuit shifts according to the intensity of the light striking the LDR: The balance can be recovered by adjusting potentiometer P1. The actual resistance of P1 is then proportional to the exposure time. The two LEDs show the balance of the bridge circuit.

To balance the circuit, adjust P1 until the two LEDS D3 and D4 turn off. After these two LEDs turn off, switch S1 to position 2. This time, capacitor C2 is connected to the supply line through S1a, R5 and P1. Capacitor C2, however, cannot charge yet because the 555 timer still shorts it to ground.

Once S2 is pressed the timer is released and the enlarger is simultaneously switched on by the relay. Photo exposure takes place until C3 charges to 2/3 of the supply voltage. Once C2 reaches this charge level, the timer resets and the relay opens thereby switching off the enlarger.

Potentiometer P2 enables the balance point of the bridge circuit to be varied in order to accomodate different exposure timings for different types of photographic paper. The setting of P2 must be found experimentally for each type of paper.

44   QUIZ REFEREE

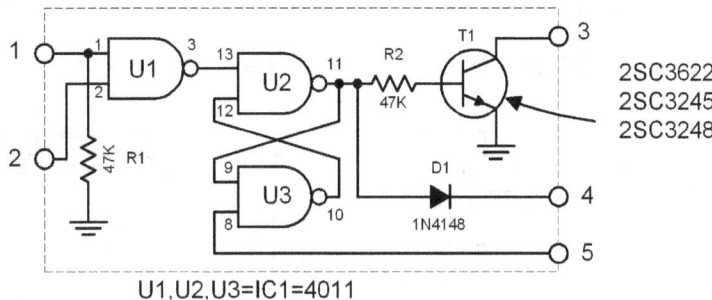

U1,U2,U3=IC1=4011

Diagram 44.0   Quiz Referee

This circuit enables a quizmaster to determine which one among the contestants believes he knows the correct answer and the first to have pressed the answer button. Once a button is pressed, its corresponding LED turns on, and all other buttons are immediately blocked out.

The circuit in the diagram 44.0 is the actual wiring of each block (SW...) in diagram 44.1. The number of buttons (and contestants) is unlimited. A duplicate block must be constructed and connected to the main circuit for every contestant. Diagram 44.1 shows how to connect several blocks.

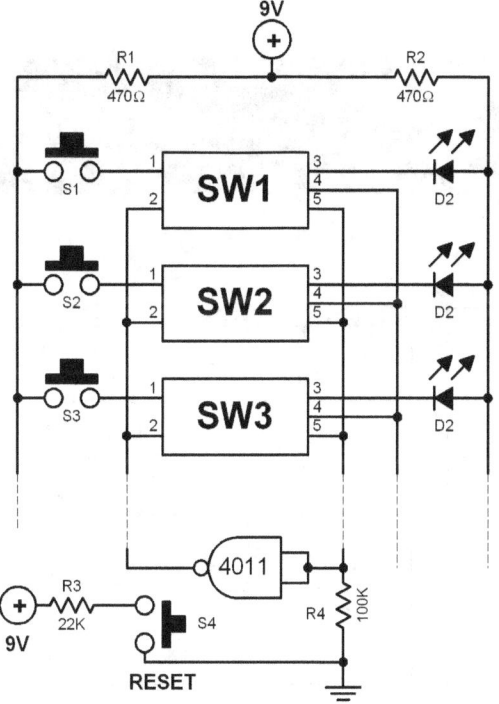

Diagram 44.1   Installation of Quiz Referee

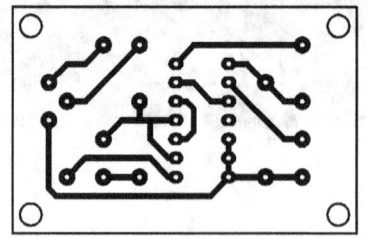

Figure 44.0 Printed Circuit Layout   **Figure 44.1** Parts Placement

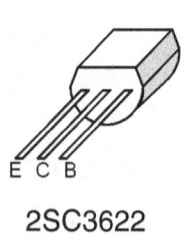

2SC3622
2SC3245
2SC3245A
2SC3248

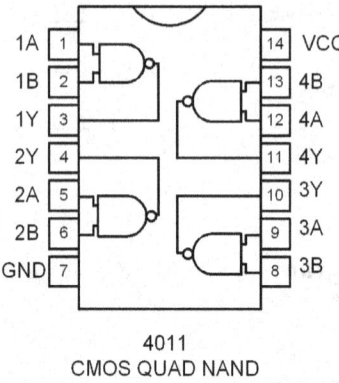

4011
CMOS QUAD NAND

Parts List:

R1,R2 = 47K
T1 = 2SC3622
      2SC3245
      2SC3248
D1 = 1N4148
IC1 = 4011

45 BIO-SIGNAL INTERFACE

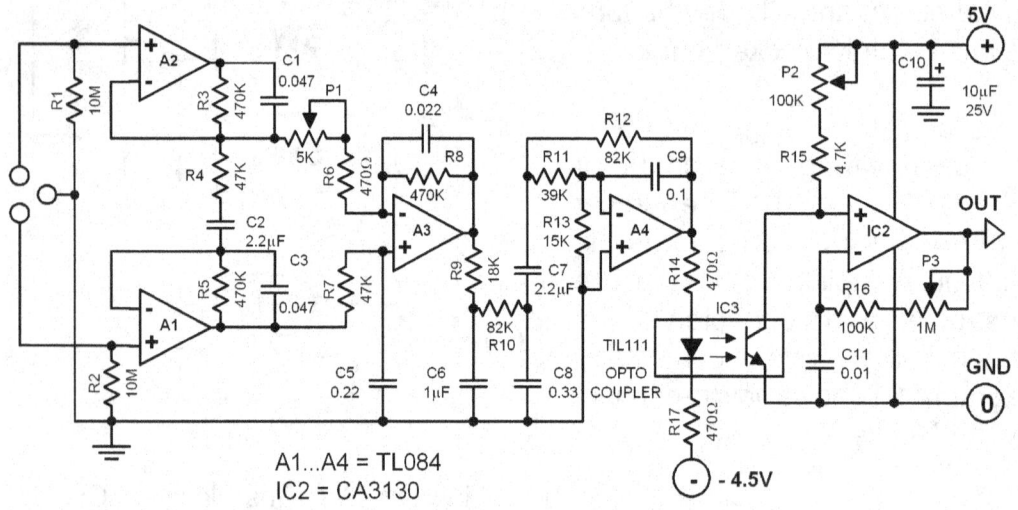

A1...A4 = TL084
IC2 = CA3130

Diagram 45.0 Bio-Signal Interface

This circuit is used as a computer interface to detect and display electrical impulses generated by the heart, the muscles, or the brain. The electrodes are made of three small copper plates which are connected to a three-wire shielded cable. The circuits composed of A1...A3 is called "instrumentation" amplifier. The amplification of the impulses is adjustable through P1. Batteries must be used to power the circuit for security reasons. Never use a supply voltage from the power lines. The author and publisher don't take responsibility for any damages that may arise from the use of the circuits.

The impulses are transmitted to the computer part of the circuit (IC2) via optocoupler. This guarantees the electrical as well as galvanic isolation of the interface. The impulse pulse width modulates a carrier signal that is generated by the IC2 itself. P2 sets the duty cycle of the carrier signal to 50% with short circuited inputs. P3 sets the carrier frequency . In order to display the electrical impulses that are detected from the human body, an appropriate computer software must be developed.

  Never use a voltage supply connected to power lines. Use batteries only in powering the interface circuit!

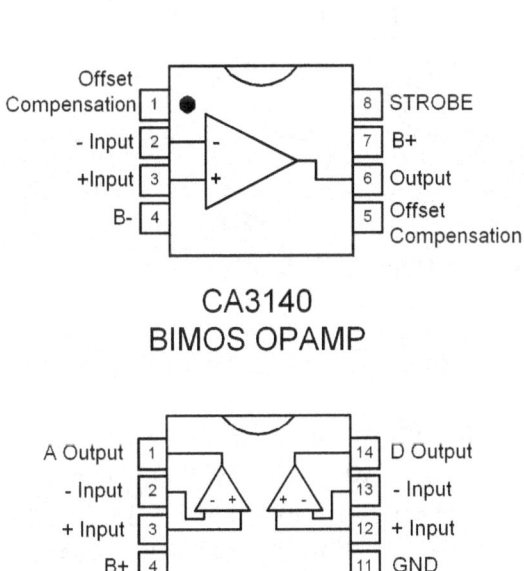

Offset Compensation 1 — 8 STROBE
- Input 2 — 7 B+
+Input 3 — 6 Output
B- 4 — 5 Offset Compensation

CA3140
BIMOS OPAMP

A Output 1 — 14 D Output
- Input 2 — 13 - Input
+ Input 3 — 12 + Input
B+ 4 — 11 GND
+ Input 5 — 10 + Input
- Input 6 — 9 - Input
B Output 7 — 8 C Output

TL084
QUAD OPAMP

Parts List:	
R1,R2 = 10M	C1,C3 = 0.0047/50V
R3,R5,R8 = 470K	C2,C7 = 2.2µF(np)
R4,R7 = 47K	C4 = 0.022/50V
R6,R14,R17 = 470Ω	C5 = 0.22/50V
R9 = 18K	C6 = 1µF(np)
R10,R12 = 82K	C8 = 0.33/50V
R11 = 39K	C9 = 0.1/50V
R13 = 15K	C10 = 10µF/25V
R15 = 4.7K	C11 = 0.01/50V
R16 = 100K	

P1 = 5K trimmer
P2 = 100K trimmer
P3 = 1M trimmer

IC1= TL084 Quad Opamp w/ JFET inputs
IC2 = CA3130 BIMOS Opamp
IC3 = Optocoupler (TIL111)

46   ONE ARM BANDIT

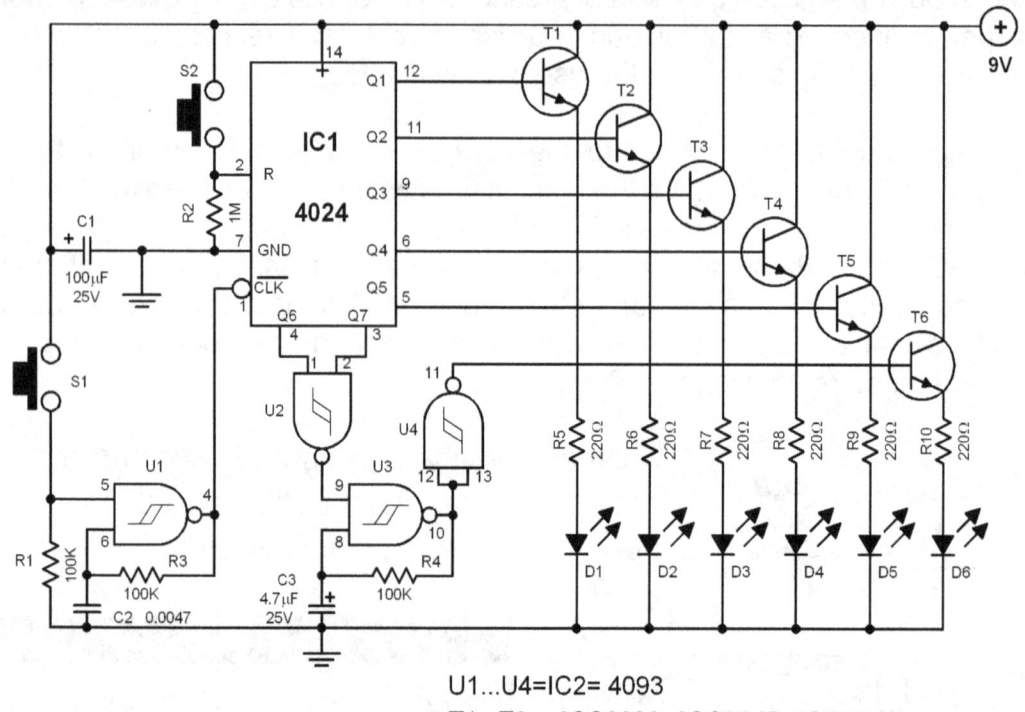

Diagram 46.0   One Arm Bandit

This is the electronic version of the one-arm-bandit game machine commonly found in casinos. With this circuit, you never need to throw coins in. However, you never win money too. You can use this game circuit however for generating random patterns. If you give each LED a unique number, you can generate random number patterns. Such random numbers can be useful for applications like encryption keywords.

The main player of the circuit is the asynchronous 7-stage binary counter (IC1= 4024). It is reset by switch S2 before every start. At this instance, all the LEDs are turned off. The "game" is started by pressing the button S1. Once this button is pressed, the clock oscillator activates and drives the counter. Once the button is released, the clock stops and the last binary count (generated at random) is displayed by the LEDs. The LEDs can be given values according to the players' desire.

The circuit consumes minimal current since the ICs are CMOS type. The consumption is mainly dependent on the type of LEDs used. The voltage supply can be between 4.5 volts and 9 volts. If the circuit is used only occasionally, a single 9 volt block battery is sufficient.

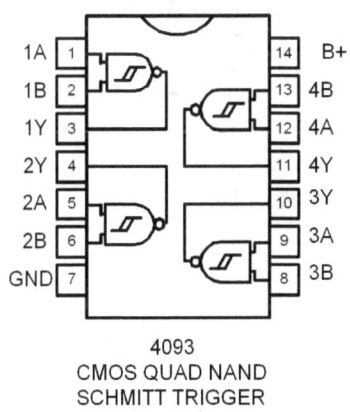

4093
CMOS QUAD NAND
SCHMITT TRIGGER

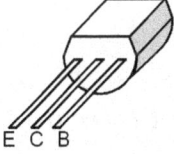

2SC3622
2SC3245
2SC3245A
2SC3248

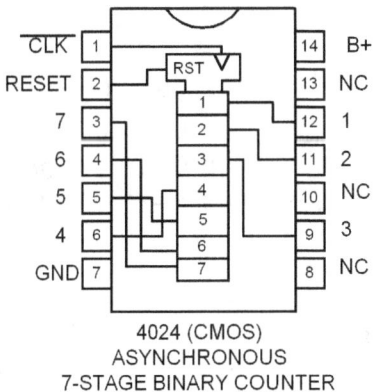

4024 (CMOS)
ASYNCHRONOUS
7-STAGE BINARY COUNTER

Parts List:

R1,R2,R3,R4= 100K
R5,R6,R7 = 220Ω
R8,R9,R10 = 220Ω
C1 = 100µF/25V
C2 = 0.0047/50V
C3 = 4.7µF/25V
D1,D2,D3,D4,D5,D6 = 1N4148
T1,T2,T3,T4,T5,T6 = 2S3622
  (2SC3245,2SC3248)
IC1= 4024
IC2 = 4093

47 GUITAR SOUND LIMITER

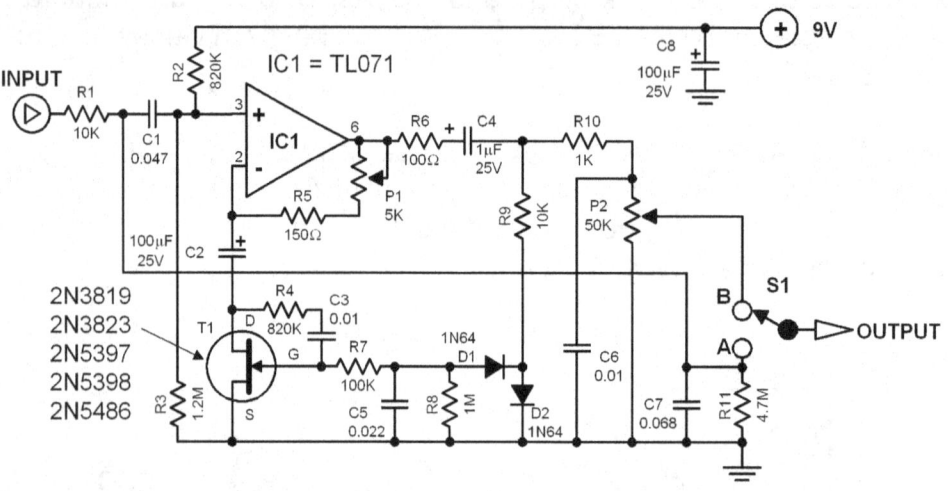

Diagram 47.0 Guitar Sound Limiter

This circuit limits the sound output of an electric guitar by compressing its signal peaks. This limiting technique is usually employed to avoid distortions caused by overdrive.

The circuit is simple. A single IC TL071 works as preamp and the FET transistor works as a voltage controlled resistor. The FET is being controlled by a negative rectified voltage sampled from the output of IC1. When the output increases, the resistance of T1 also increases and reduces the amplification of IC1.

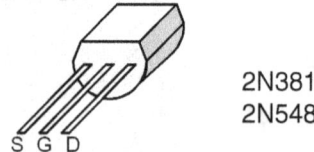

2N3819
2N5486

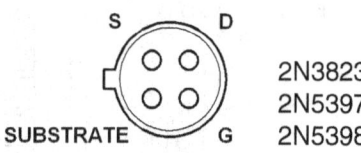

2N3823
2N5397
2N5398

Calibration: You need an oscilloscope and a signal generator. A 1 kHZ signal of 150 mV is fed into the input and P1 is adjusted to the maximum amplification without distortion as seen in the oscilloscope. Then, the input is increased to 300 mV and P1 is adjusted back until the distortion is reduced to an acceptable degree.

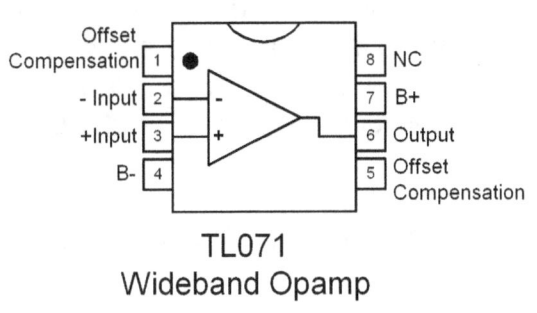

TL071
Wideband Opamp

Parts List:

R1,R9 = 10K	C1 = 0.047/50V
R2,R4 = 820K	C2,C8 = 100µF/25V
R3 = 1.2M	C3,C6 = 0.01/50V
R5 = 150Ω	C4 = 1µF/25V
R6 = 100Ω	C5 = 0.022/50V
R7 = 100K	C7 = 0.068/50V
R8 = 1M	D1,D2= 1N64
R10 =1K	T1 = 2N3819(2N5486)
R11 = 4.7M	IC1= TL071

48 PHOTO FIXER TIMER

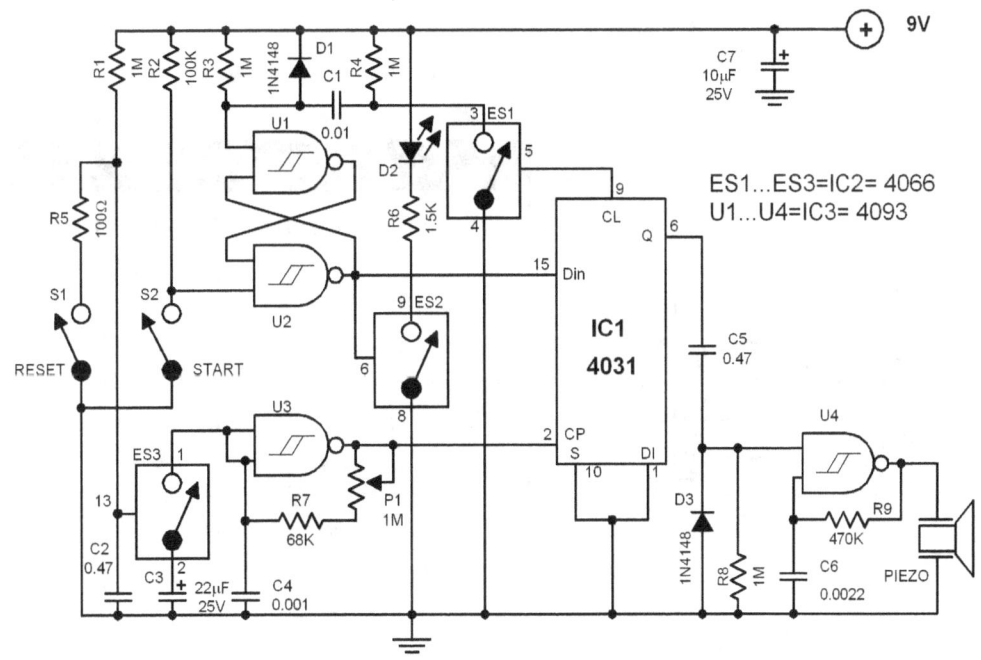

ES1...ES3=IC2= 4066
U1...U4=IC3= 4093

Diagram 48.0 Photo Fixer Timer

This timer is used in photographic reproduction work to time how long must the photographic paper remain in the fixer solution. The operation of the circuit is simple. When a photographic paper is placed in the fixer solution, the button S2 is pressed and an LED lights as an acknowledgement. When the desired fixing time has elapsed, the circuit beeps signalling that it is time to fish the photo out of the fixer solution.

The fixing time can be varied through P1 between 1 minute and 10 minutes. The button S1 serves as the reset.

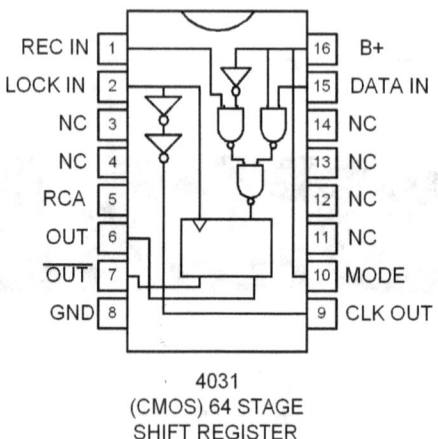

4031
(CMOS) 64 STAGE
SHIFT REGISTER

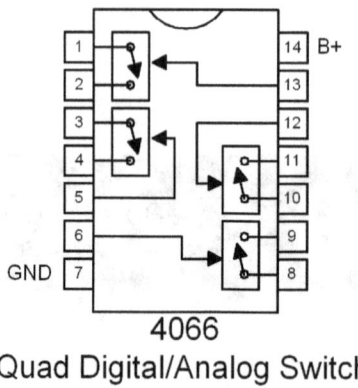

4066
Quad Digital/Analog Switch

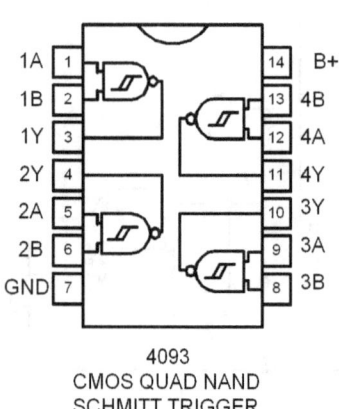

4093
CMOS QUAD NAND
SCHMITT TRIGGER

Parts List:	
R1,R3,R4,R8 = 1M	D1,D3 = 1N4148
R2 = 100K	D2 = LED
R5 = 100Ω	
R6 = 1.5K	IC1 = 4031
R7 = 68K	IC2 = 4066
R9 = 470K	IC3 = 4093
C1 = 0.01/50V	
C2,C5 = 0.47/50V	Piezo transducer
C3 = 22µF/25V	Pushbutton
C4 = 0.001/50V	9V battery
C6 = 0.0022/50V	
C7 = 10µF/25V	

49 LED BLINKER

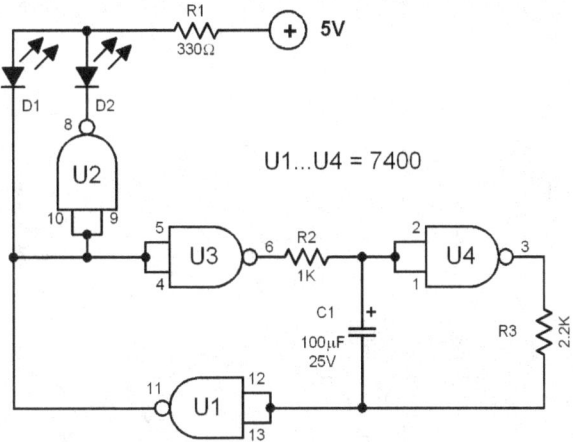

Diagram 49.0   LED Blinker

This blinker circuit is universal. It can be used in model railways, small ads, modelcars, etc. -practically everything that needs a nonstatic optical signal. The components around A1 determine the frequency of the blinking. A potentiometer can be used in place of R3 to enable easy variation of the blink frequency.

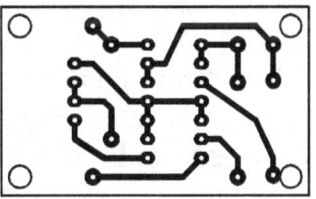

Figure 49.0   Printed Circuit Layout

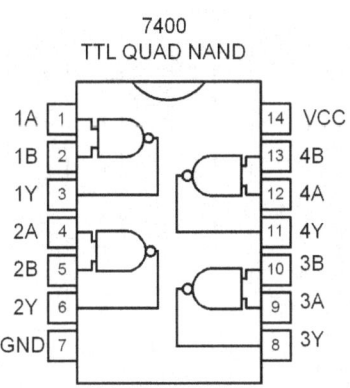

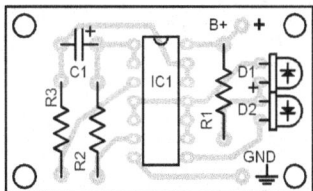

Figure 49.1   Parts Placement

50 SHIP HORN CONTROL

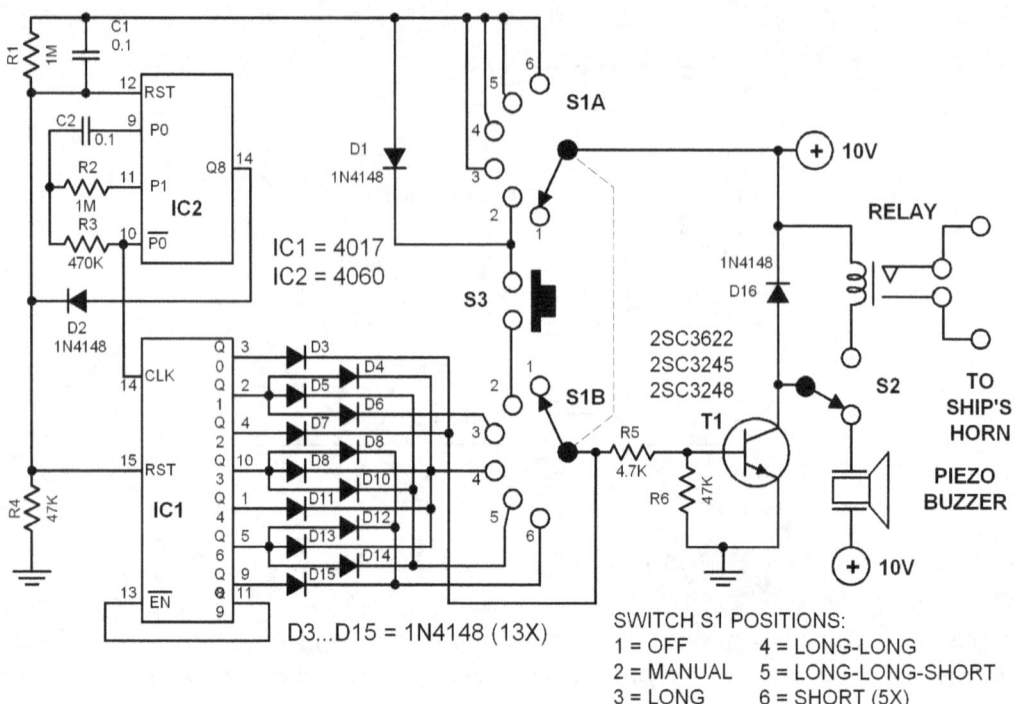

IC1 = 4017
IC2 = 4060

D3...D15 = 1N4148 (13X)

SWITCH S1 POSITIONS:

1 = OFF	4 = LONG-LONG
2 = MANUAL	5 = LONG-LONG-SHORT
3 = LONG	6 = SHORT (5X)

Diagram 50.0   Ship Horn Control

This circuit automatically controls a ship's horn during heavy fog. Four tone combinations are selectable through S1. The tone combinations are the following:

3) single long tone
4) two long tones
5) one long-short-short
6) five short tones

The selected tone combination is blown every 2 minutes.

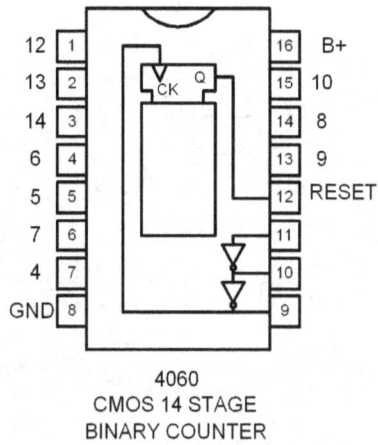

4060
CMOS 14 STAGE
BINARY COUNTER

51 STEPPER MOTOR CONTROL

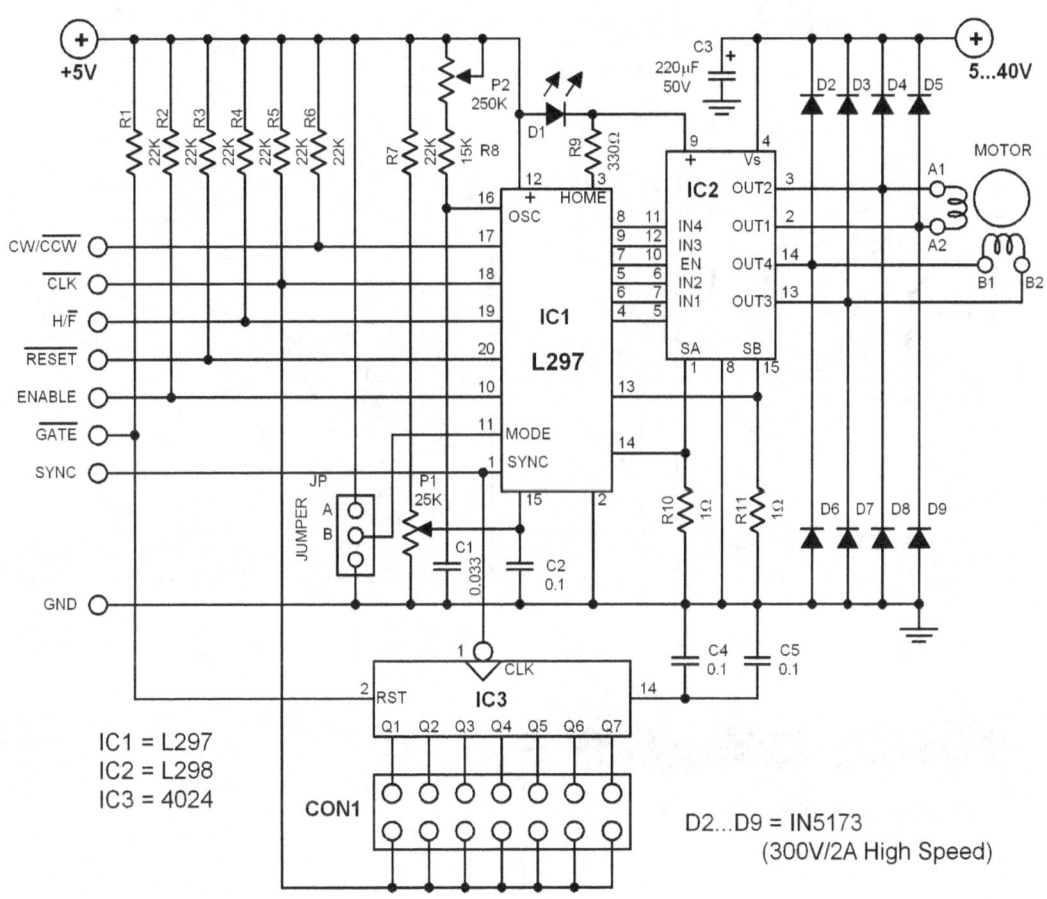

Diagram 51.0 Stepper Motor Control

This circuit is designed to control a double stator stepper motor. A double stator motor has either 2 bipolar coils or 4 unipolar coils. The circuit is built around the IC pair L297 and L298. The controller has a common computer I/O connection, an integrated motor controller and driver stages, and an optional clock generator.

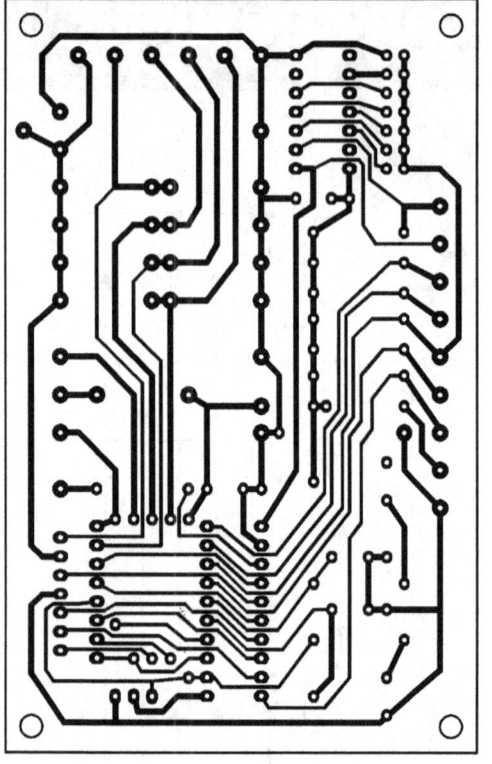

Figure 51.0 Printed Circuit Layout   **Figure 51.1** Parts Placement

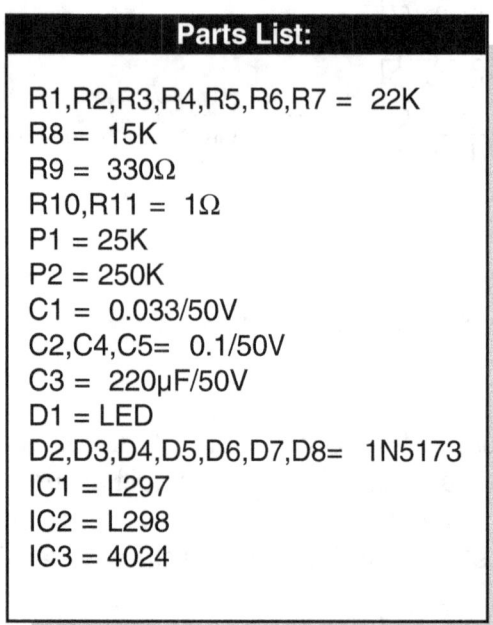

Parts List:

R1,R2,R3,R4,R5,R6,R7 = 22K
R8 = 15K
R9 = 330Ω
R10,R11 = 1Ω
P1 = 25K
P2 = 250K
C1 = 0.033/50V
C2,C4,C5= 0.1/50V
C3 = 220µF/50V
D1 = LED
D2,D3,D4,D5,D6,D7,D8= 1N5173
IC1 = L297
IC2 = L298
IC3 = 4024

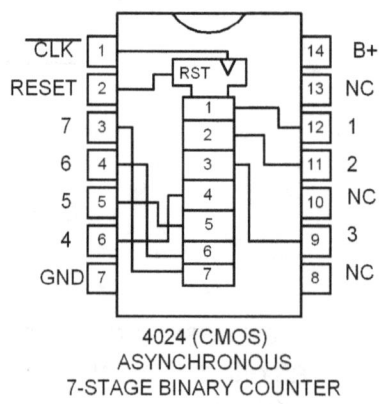

4024 (CMOS)
ASYNCHRONOUS
7-STAGE BINARY COUNTER

52 CHESS TIMER

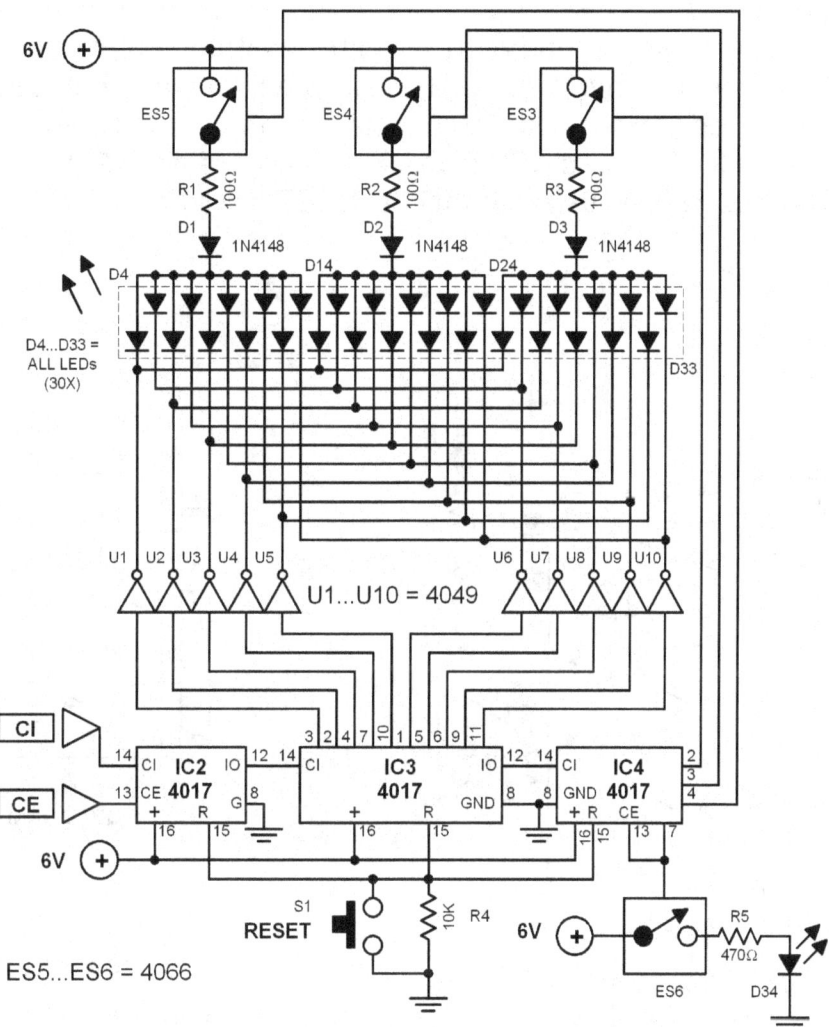

Diagram 52.0   Chess Timer

This electronic chess clock is especially designed for the so-called fast chess play where each of the player has only a maximum of 5 or 10 minutes to think before making his next move. The circuit has a counter for each player. A counter starts to count every time the other player has made his next move and touched his sensor switch.

Electronic Circuits - 1.1

Within the 5-minute time allowance the LEDs light one after the other in 10 seconds interval. Within the 10-minute time allowance, the interval is 20 seconds. The total time allowance is selectable through S2. Once the time limit is exceeded, LED D34 lights up. LEDs D39 and D40 show which player is being timed.

At the start of every game, S1 must be pressed to reset all the counters. The LEDs are constructed in a form of a circle. To calibrate the timer, an accurate stopwatch must be used as reference. Potentiometer P2 must be set so that each LED will light for exactly 10 seconds (S2 in position 1). After setting P2, S2 is switched to position 2 and P1 is adjusted so that each LED will light for 20 seconds.

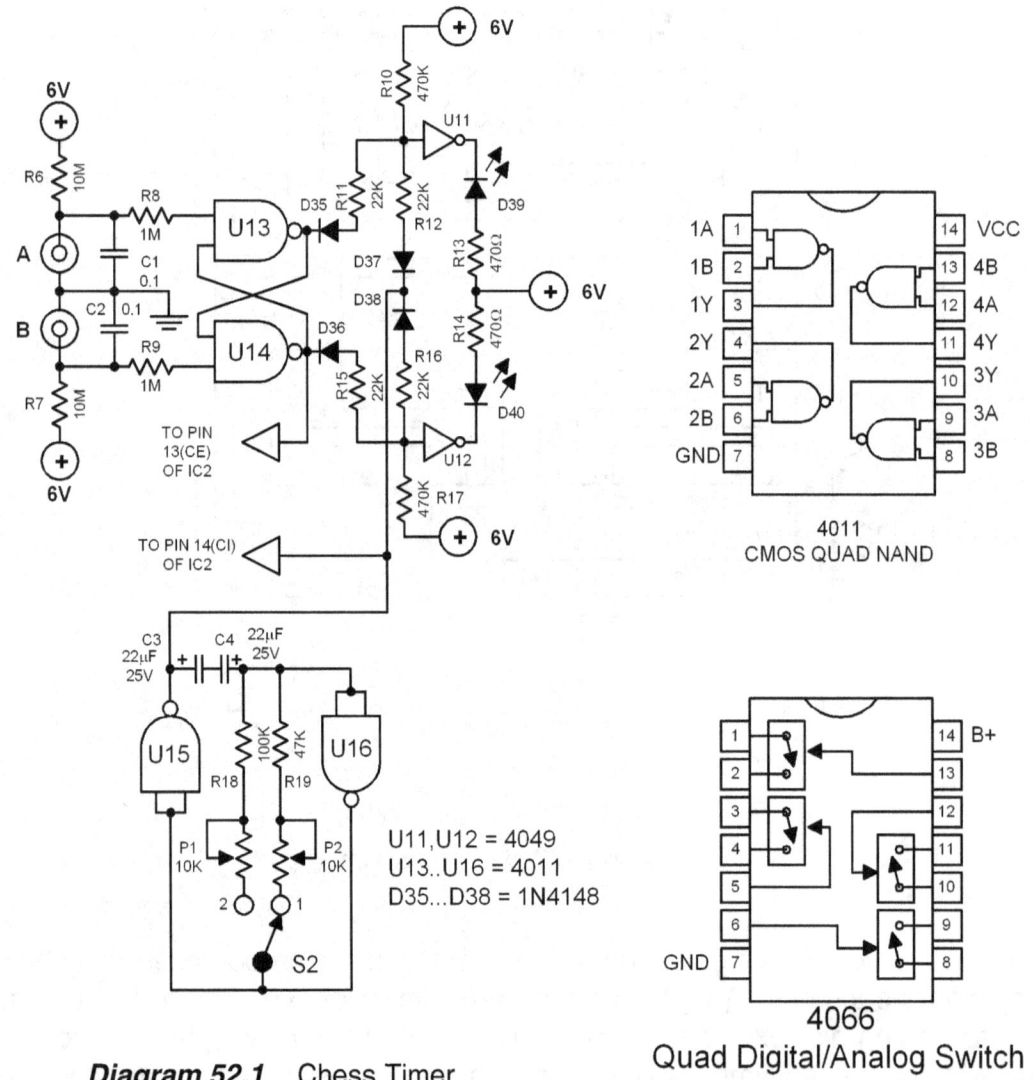

Diagram 52.1 Chess Timer

U11,U12 = 4049
U13..U16 = 4011
D35...D38 = 1N4148

4011
CMOS QUAD NAND

4066
Quad Digital/Analog Switch

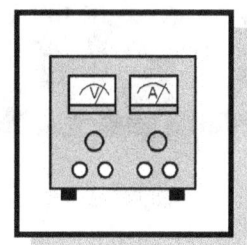

POWER SUPPLIES & CHARGERS

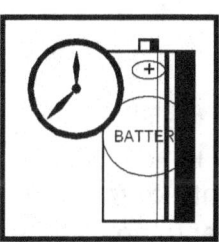

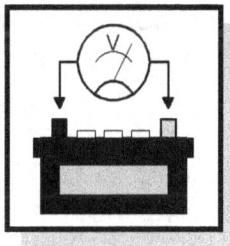

53 STABLE POWER SUPPLY

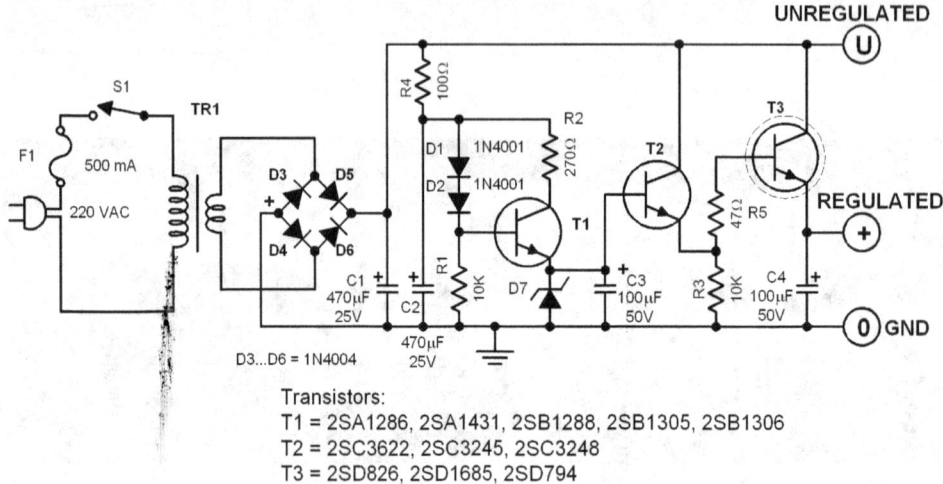

Transistors:
T1 = 2SA1286, 2SA1431, 2SB1288, 2SB1305, 2SB1306
T2 = 2SC3622, 2SC3245, 2SC3248
T3 = 2SD826, 2SD1685, 2SD794

Diagram 53.0 Stable Power Supply

The excellent voltage regulation of this circuit is achieved by feeding a constant current to the zener diode D7. The constant current source is composed of T1,D2,R1,R2. The output voltage can be selected between 3 volts and 47 volts. The output voltage is equal to Vz+2(0.7V). Vz is the zener voltage. The value 2(0.7V) comes from the base-emitter voltage of the two transistors T2 and T3. These two transistors work together as emitter follower.

When the circuit is to be used for devices which consume currents higher than 1 ampere, a 2N3055 transistor must be added to it as shown in Diagram 53.1. Transistors T3 and eventually the additional transistor T4 must be properly heatsinked.

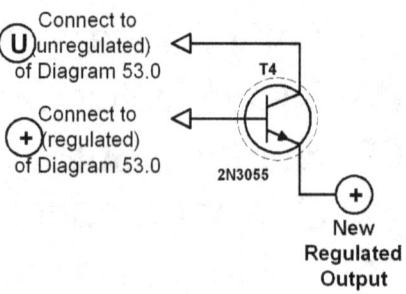

Diagram 53.1 Additional 2N3055

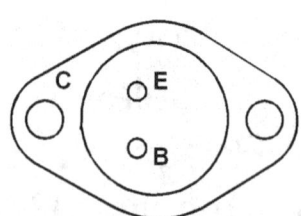

Bottom view
2N3055

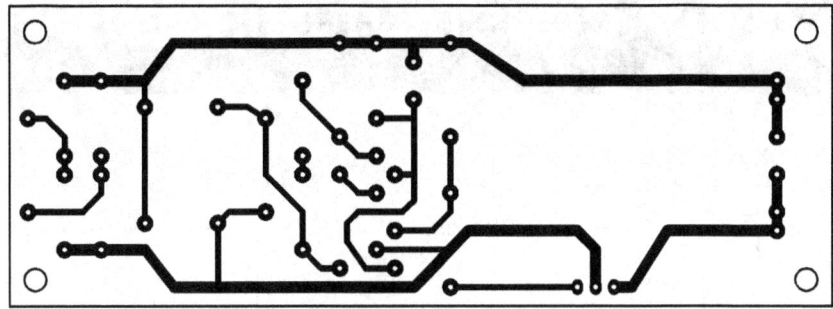

Figure 53.0 Printed Circuit Layout

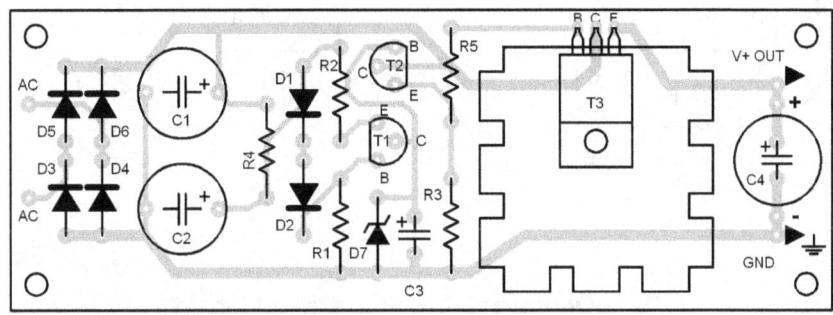

Figure 53.1 Parts Placement

Parts List:

R1,R3 = 10K
R2 = 270Ω
R4 = 100Ω
R5 = 47Ω
C1,C2 = 470µF/25V
C3,C4 = 100µF/25V
D1,D2 = 1N4001
D3,D4,D5,D6= 1N4004
D7 = Zener diode(see text)
T1 = 2SA1286
T2 = 2SC3622
T3 = 2SD826

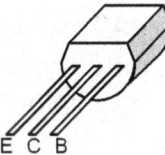

E C B

2SC3622   2SA1286
2SC3245   2SB1288
2SC3245A   2SB1305
2SC3248   2SB1306

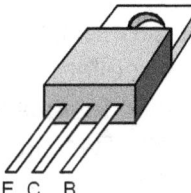

E C B

2SD826
2SD794

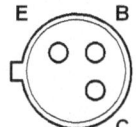

E   B

C

2SA143

E C B

2SD1682
2SD1685

54 CONSTANT CURRENT

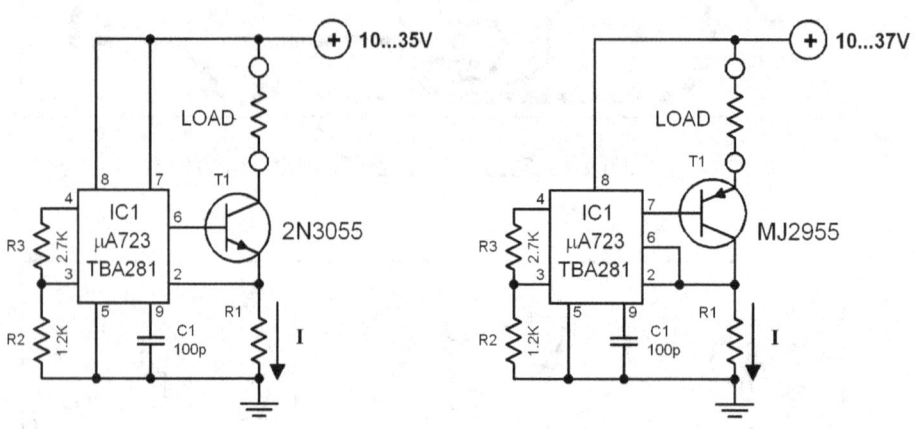

Diagram 54.0   Constant Current

There are different ways to construct a constant current source using a voltage regulator. This time we use the μA723. This IC is sometimes labeled as TBA281. The excellent stability and temperature coefficient of this IC is well known. It is actually designed as a voltage regulator but can be applied to regulate current as well. Two circuit designs are featured here - one which uses a 2N3055 power transistor and another which uses a MJ2955 transistor. The constant current I can be found with this formula:

$$I = \frac{2.2V}{R1}$$

The load is connected to pin 7 of the IC. The maximum current load that can be handled safely by the IC is around 150 mA. Add a high power transistor to the circuit to increase it current handling capacity. Circuit 1 of Diagram 54.0 uses an NPN transistor while circuit 2 uses a PNP transistor.

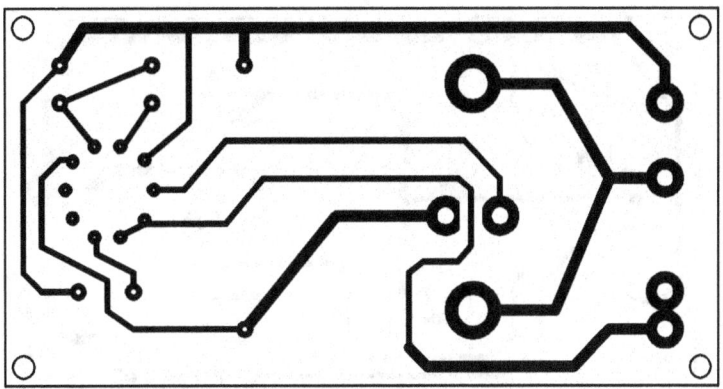

Figure 54.0 Printed Circuit Layout for
Diagram 54.0 (w/ NPN transistor)

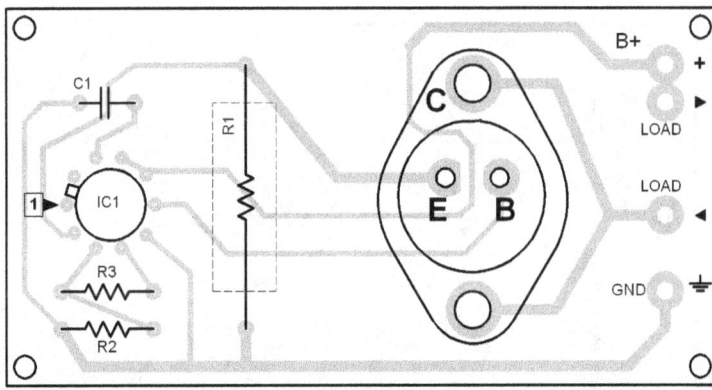

Figure 54.1 Parts Placement Layout for
Diagram 54.0 (w/ NPN transistor)

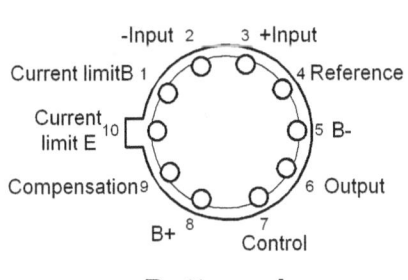

Bottom view
μA723

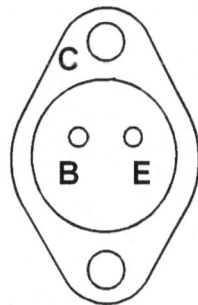

Bottom view
MJ2955

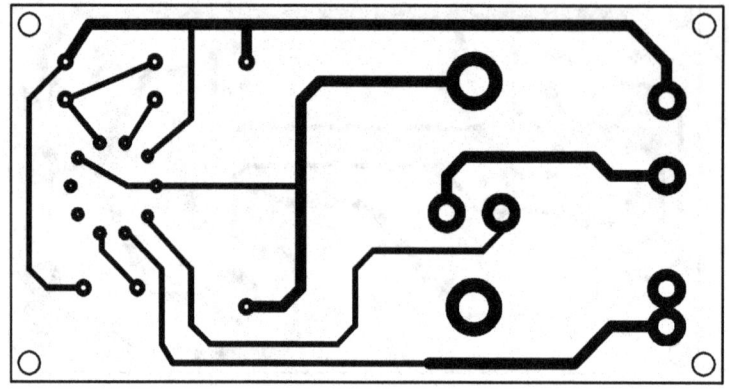

Figure 54.2 Printed Circuit Layout for Diagram 54.1 (w/ PNP transistor)

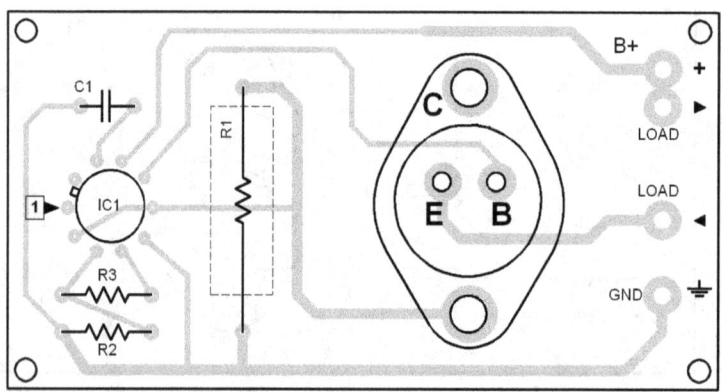

Figure 54.3 Parts Placement Layout for Diagram 54.1 (w/ PNP transistor)

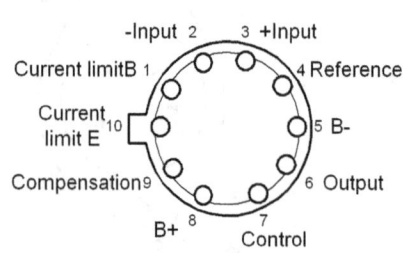

Bottom view
µA723

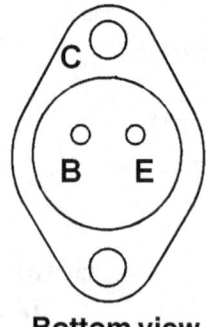

Bottom view
MJ2955

55   2N3055 DARLINGTONS

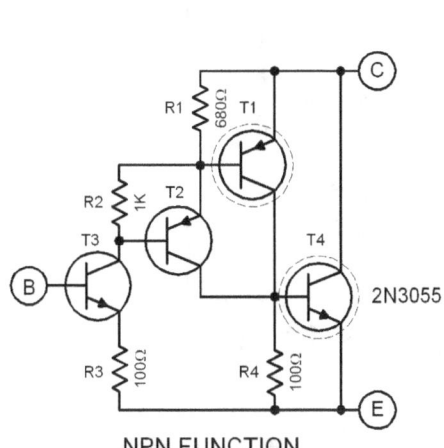

NPN FUNCTION

Transistors:
T1 = 2SB874, 2SB1144
T2 = 2SA1285, 2SA1285A
T3 = 2SC3245, 2SC3248

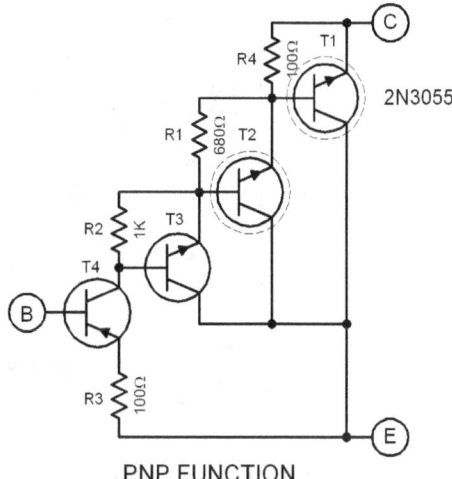

PNP FUNCTION

Transistors:
T2 = 2SD781, 2SD1177, 2SD1684
T3 = 2SC3245, 2SC3248
T4 = 2SA1285, 2SA1285A

Diagram 55.0   2N3055 Darlingtons

Very high capacity is the business of these circuits. The featured transistor combinations can be used for power supply regulators or final amplifiers that require high collector voltage, high current, high power dissipation capacity, and high current amplification characteristics. Both circuits use a 2N3055 transistor for the NPN and the PNP functions respectively. Using an NPN transistor for the PNP function may seem unusual if not incorrect to you at first glance. Through the special configuration, however, the second circuit functions as a PNP transistor. Just consider the whole circuit as a single high capacity transistor. In fact, the three solder points are labeled E,B,C. They correspond to the normal terminals of a single transistor.

The gain of the circuits (both AC and DC) is around 1.5 million. The maximum power dissipation at 25°C is 115 watts. The maximum collector voltage is 60 volts and the maximum collector current is 15 amperes. The voltage drop at the NPN combination is around 2 volts. The voltage drop at the the PNP combination is around 3 volts.

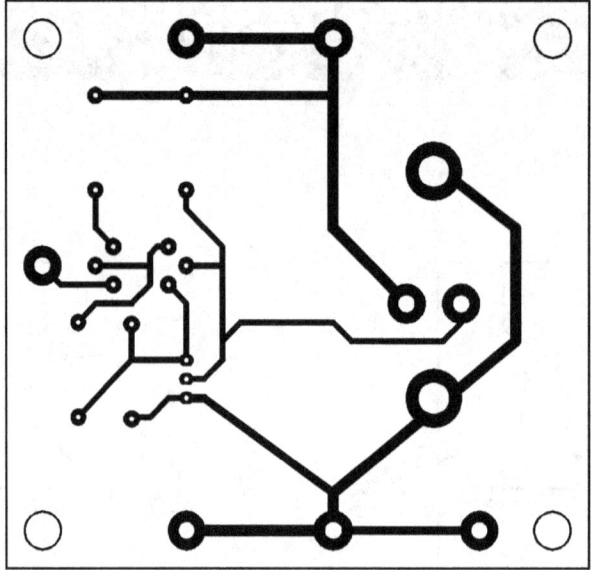

Figure 55.0 Printed Circuit Layout for the NPN function

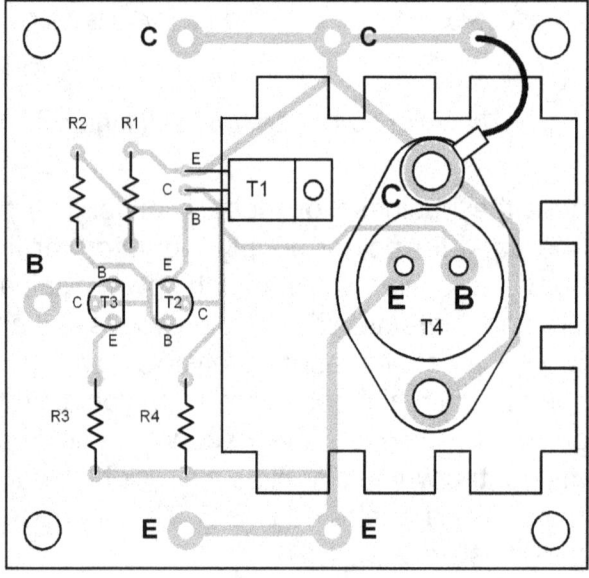

Figure 55.1 Parts Placement Layout for the NPN function

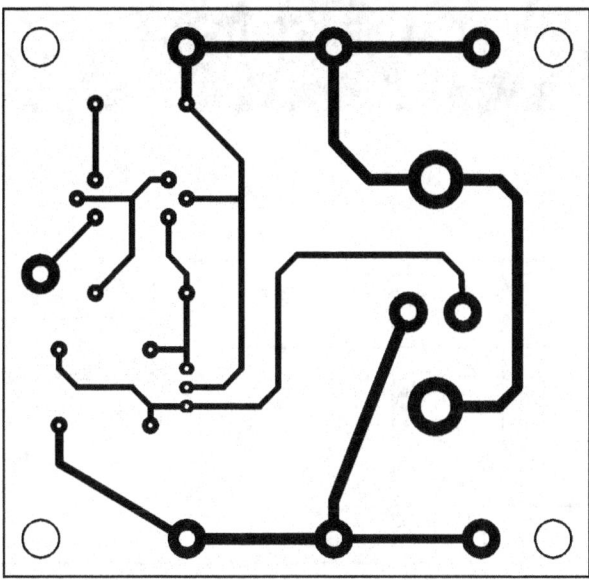

Figure 55.2 Printed Circuit Layout for the PNP function

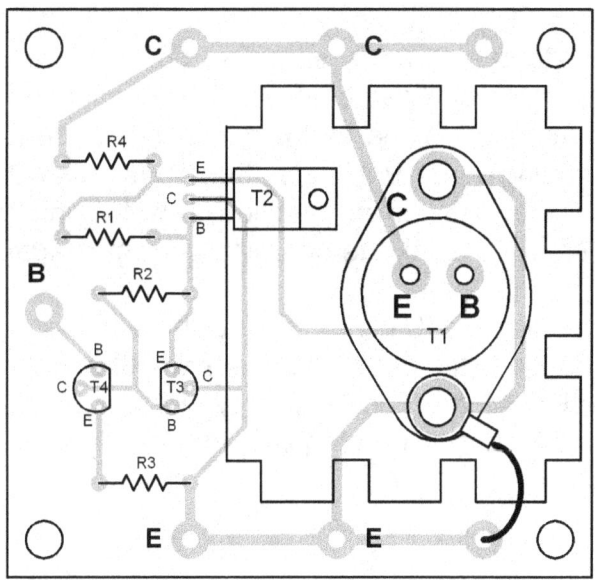

Figure 55.3 Parts Placement Layout for the PNP function

56 NICAD QUICK CHARGER

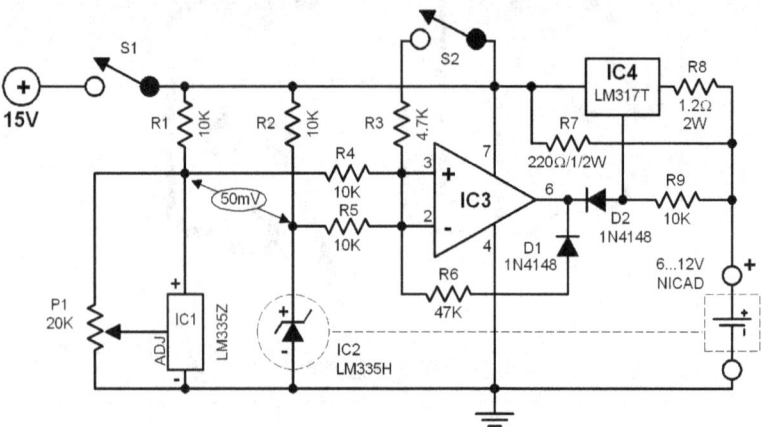

IC3 = LM308, CA3140, CA3130

Diagram 56.0 Nicad Quick Charger

This circuit uses two temperature sensing ICs from National Semiconductors to control the quick charging of Nicad baterries. The quick charging process goes this way: first the batteries are charged very fastly (high current) within the first 25% of the total charging period until the battery voltage has increased by about 1.42 volts. Then the charging current is pulled down to minimum. Shortly before reaching the fully-charged level, the charging current is again increased.

The two temperature sensors IC1 & IC2 control the charging tempo according to the temperature change in the batteries. That's why the metal case of IC2 must be thermally coupled to the Nicad battery. When the Nicad temperature increases by around 5°C, the charging is automatically shut off.

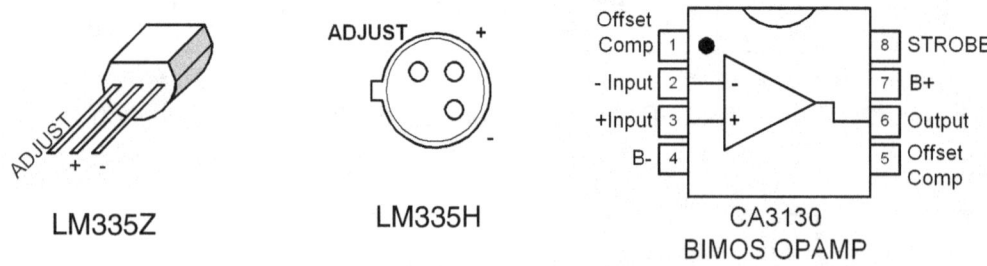

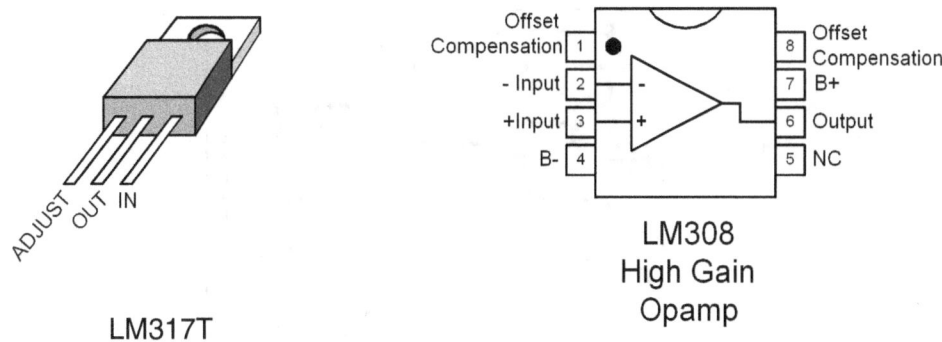

LM317T

LM308
High Gain
Opamp

57 STANDBY SUPPLY

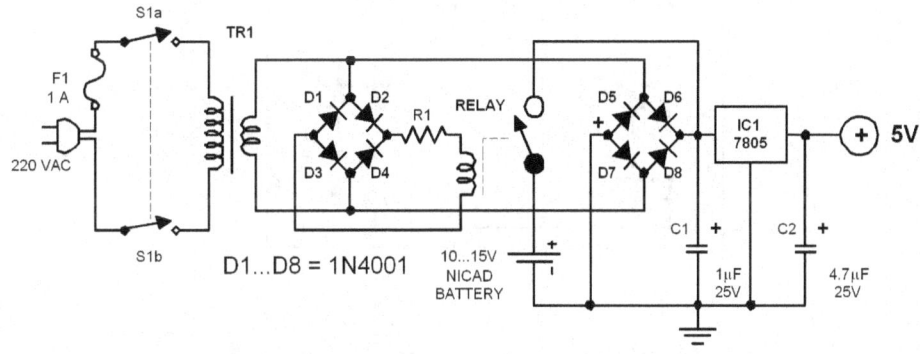

Diagram 57.0 Standby Supply

This circuit provides continous power supply for memory circuits during sudden power break downs. When the main line loses power, the relay's magnetic field collapses. Since the contact is a normally-closed type, it connects the standby battery cells to the regulator circuit thereby providing continuous power.

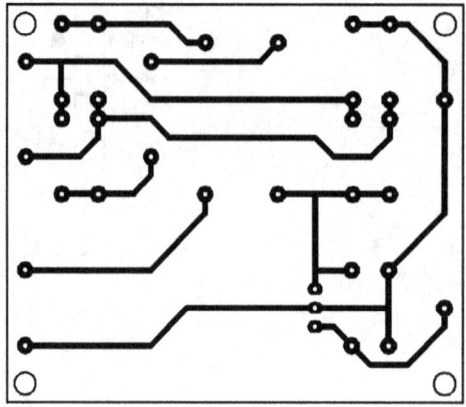

Figure 57.0 Printed Circuit Layout

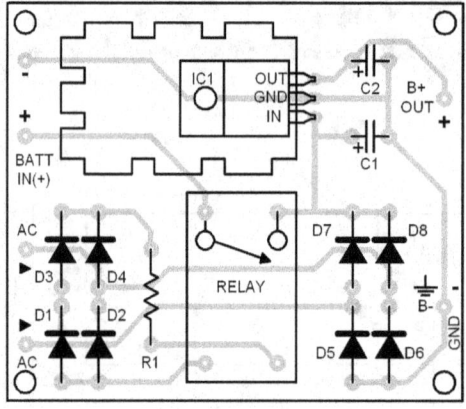

Figure 57.1 Parts Placement Layout

**7805
5-Volts
voltage regulator**

58 CURRENT MONITORED SUPPLY

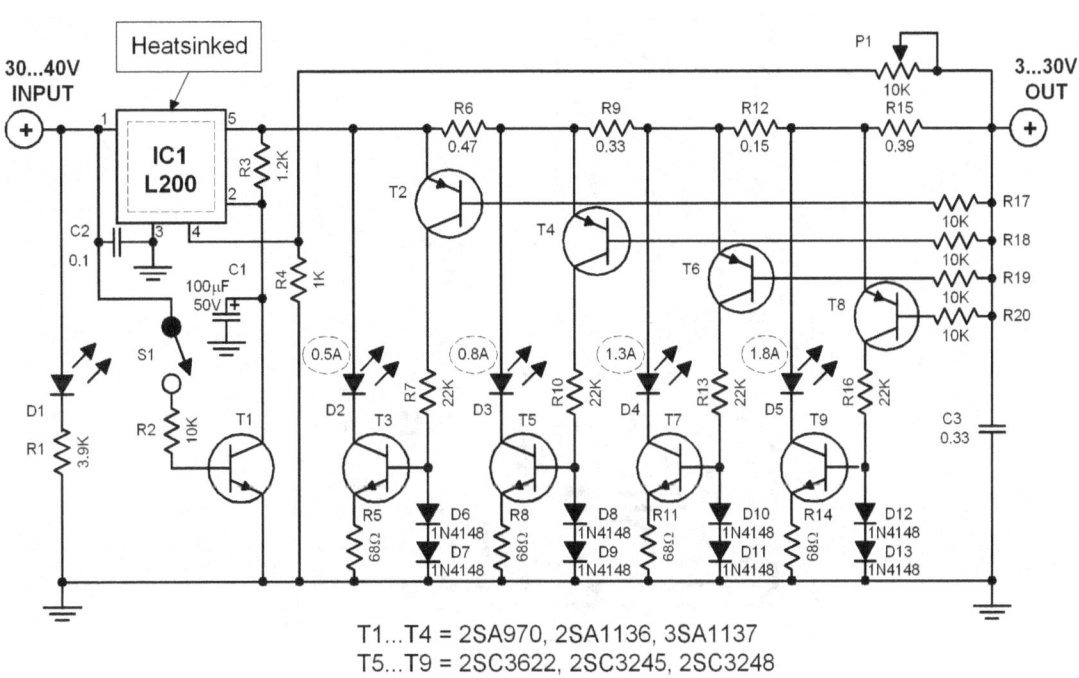

Diagram 58.0 Current Monitored Supply

Here is a reliable power that indicates the actual current being consumed by the load. Voltage regulation is done by the IC L200. Current monitoring is indicated by the LEDs D2...D5 representing current values of 0.5A, 0.8A, 1.3A, and 1.8A. The circuit can supply currents up to 2 A and its output voltage is variable from 3 volts up to 30 volts.

Normally, LED5 is red colored to serve as a warning signal that the maximum current is reached. The output voltage can be varied through P1. The circuit has a power-on delay composed of R3 and C1. Transistor T1 is the emergency shut-off.

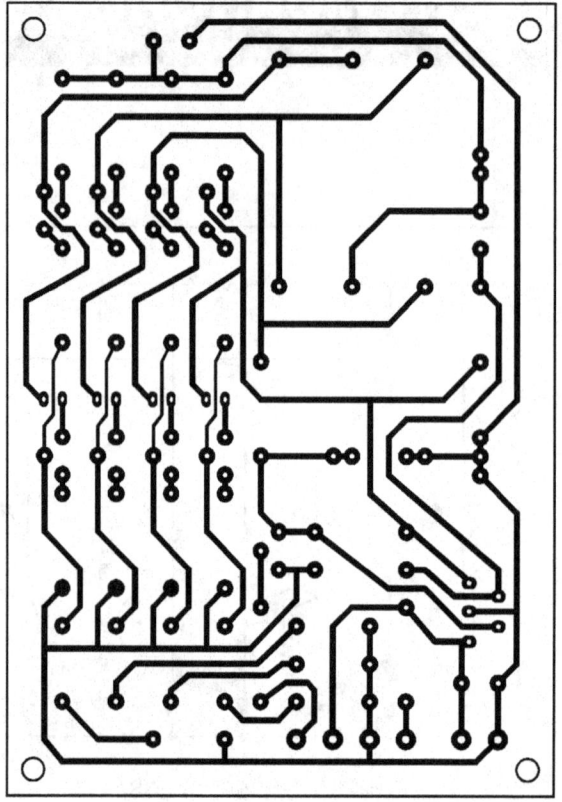

Figure 58.0   Printed Circuit Layout

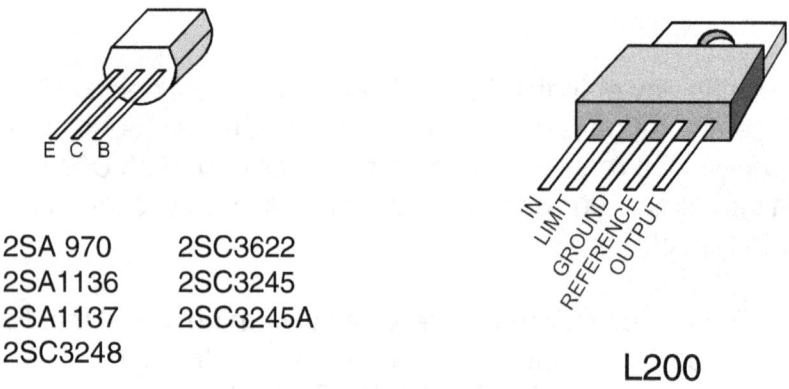

E C B

2SA 970	2SC3622
2SA1136	2SC3245
2SA1137	2SC3245A
2SC3248	

IN LIMIT GROUND REFERENCE OUTPUT

L200

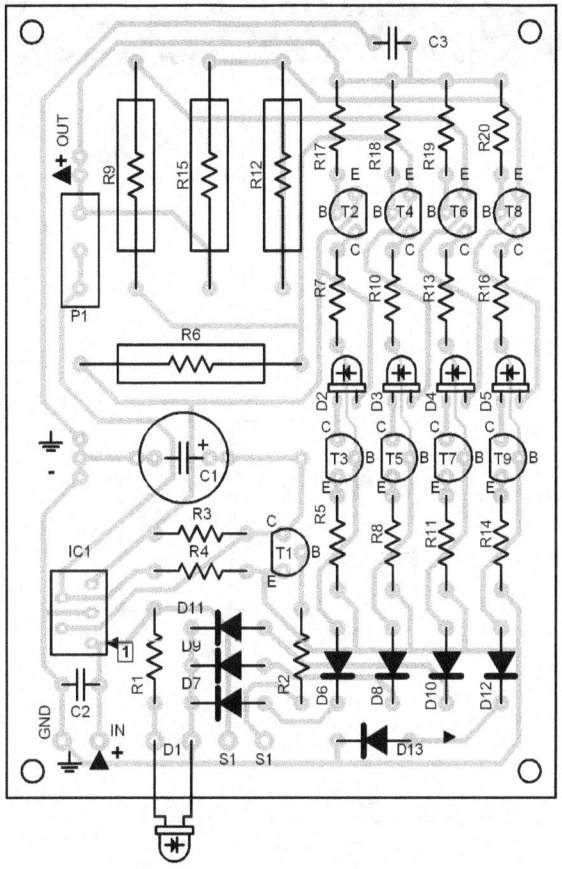

Figure 58.1   Parts Placement Layout

Parts List:

R1 = 3.9K
R2,R17,R18,R19,R20 = 10KΩ
R3 = 1.2K
R4 = 1K
R5,R8,R11,R14 = 68Ω
R6 = 0.47Ω/5W
R7,R10,R13,R16 = 22K
R9 = 0.33Ω/5W
R12 = 0.15Ω/5W
R15 = 0.39Ω/5W

C1 = 100µF/50V
C2 = 0.1/50V
C3 = 0.33/50V
D1,D2,D3,D4,D5 = LED
D6,D7,D8,D9 = 1N4148
D10,D11,D12,D13 = 1N4148
T1,T3,T5,T7,T9 = 2SC3622
  (2SC3245)(2SC3248)
T2,T4,T6,T8 = 2SA970
  (2SA1136)(2SA1137)
IC1 = L200

59 POLARITY INVERTER

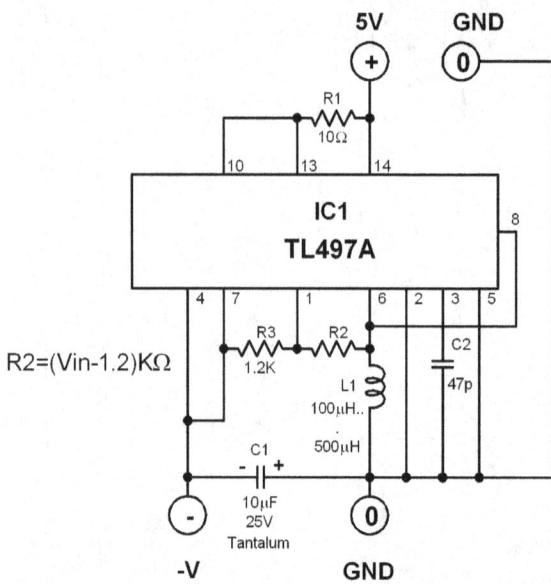

Diagram 59.0 Polarity Inverter

This circuit is used to reverse the polarity of a supply line. A negative potential is sometimes necessary in computer or battery powered circuits. Since the current consumption for such a need is mostly low, a single IC can be used as a polarity converter. The TL497A IC is originally designed for downward or upward transformation applications but is used here as polarity converter. The design does not need any transformer or bridge rectifiers.

L1 can be between 100 and 500 µH. The output voltage (Vout) can be found with this formula:

$$V_{out} = \frac{-1.2}{R2}$$

 **The load current must not exceed 500 mA.**

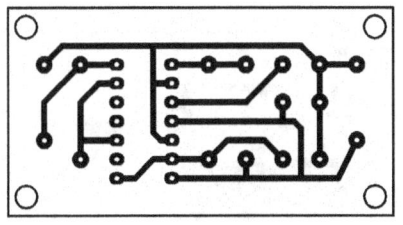

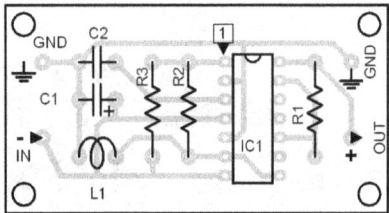

Figure 59.0   Printed Circuit Layout   ***Figure 59.1***   Parts Placement Layout

60   SUPPLY FOR OPAMPS

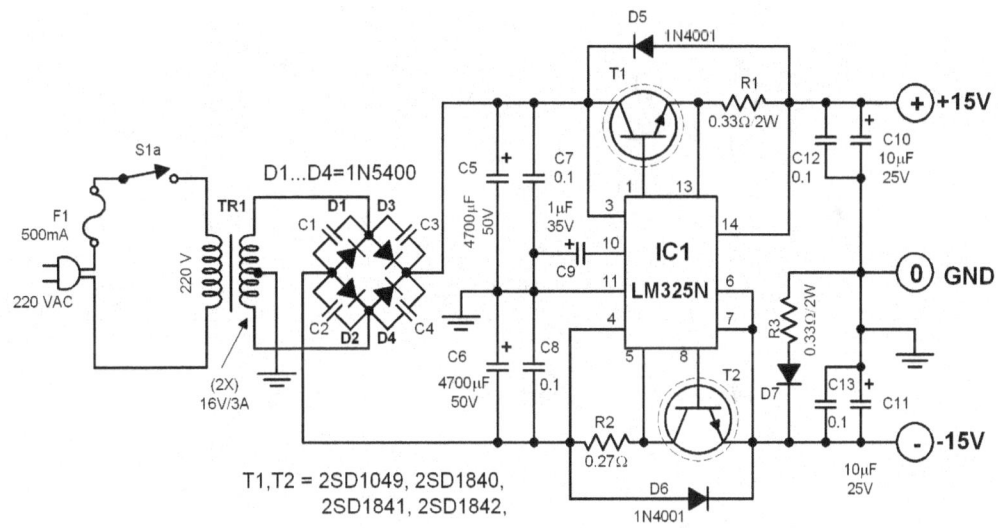

Diagram 60.0   Supply for OPAMPS

This power supply circuit is designed to support a large number of opamp ICs like in mixer circuits and sound synthesizers. The regulation is done by the IC LM325. The output voltage is +/- 15 volts symmetrical. The output current can be up to 2 amperes. Capacitors C9, C10, and C11 are tantalum types. The two power transistors T1 and T2 must be properly heatsinked.

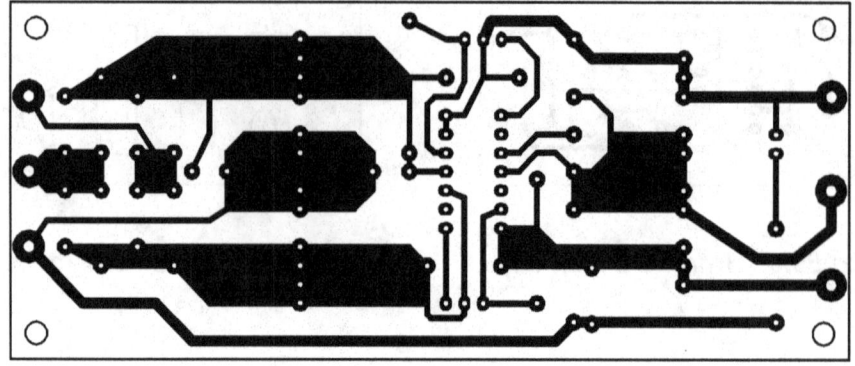

Figure 60.0   Printed Circuit Layout

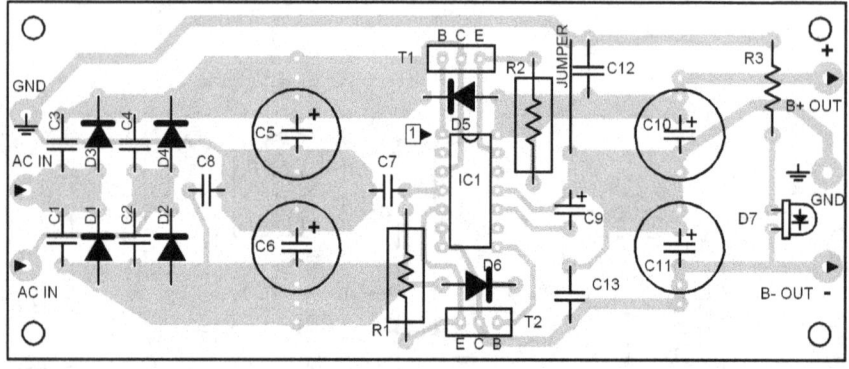

Figure 60.1   Parts Placement Layout

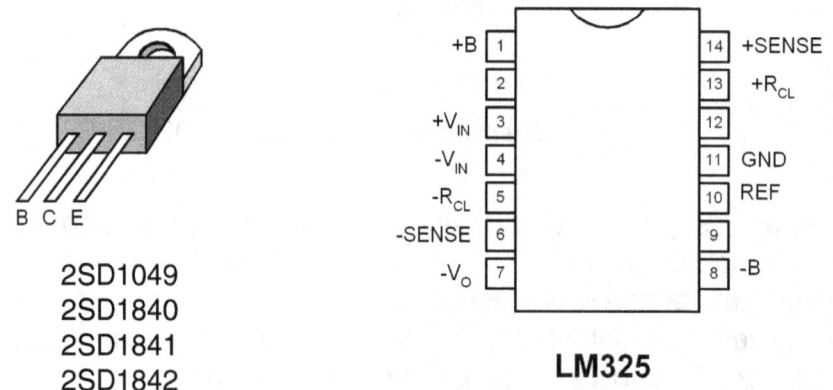

61 COMPUTER CONTROLLED POWER REGULATION

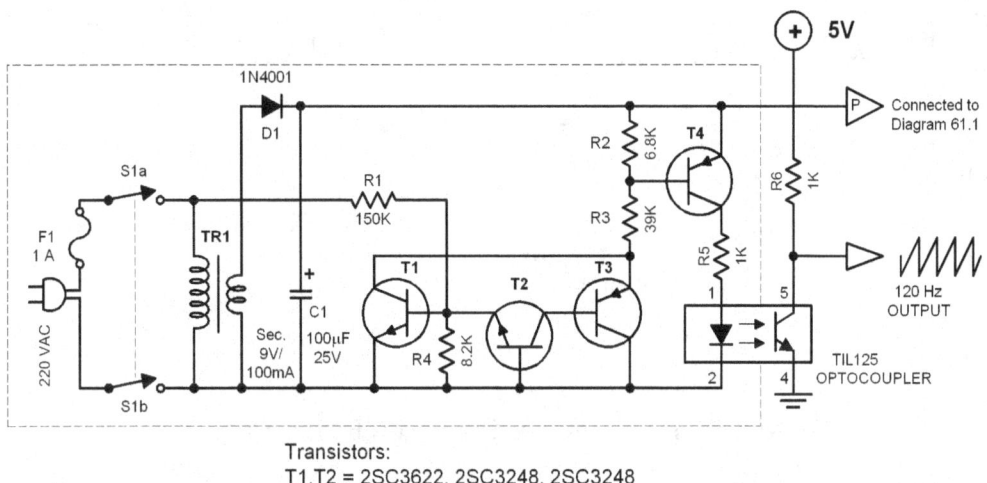

Transistors:
T1,T2 = 2SC3622, 2SC3248, 2SC3248
T3,T4 = 2SA970, 2SA1136, 2SA1137

Diagram 61.0 Computer controlled power regulation

This computer interface can control an electrical load like a lamp, electric drill, etc. in 255 steps. The maximum power handling capacity of the circuit is 400 watts. The power regulation is done by controlling the voltage level at the connected load (see Diagram 61.1).

Optocouplers are used to transmit the computer control to the power line part of the circuit. With this technique, the computer is electrically and galvanically isolated from the main power control circuit. This isolation is very important to protect the computer from possible damages caused by high voltage peaks. The phase angle of the voltage that triggers the triac is dependent on the value of the byte delivered by the computer. The higher the byte value, the higher the phase angle becomes, and the shorter will be the time when the triac remains conducting.

Calibration: Turn P1 while the input of the counter sees FFhex (255dec) until the voltage output is 0. The data 00hex produces the same output as the data FFhex. Full power is delivered by data 01hex. The triac must be rated twice the maximum load consumption.

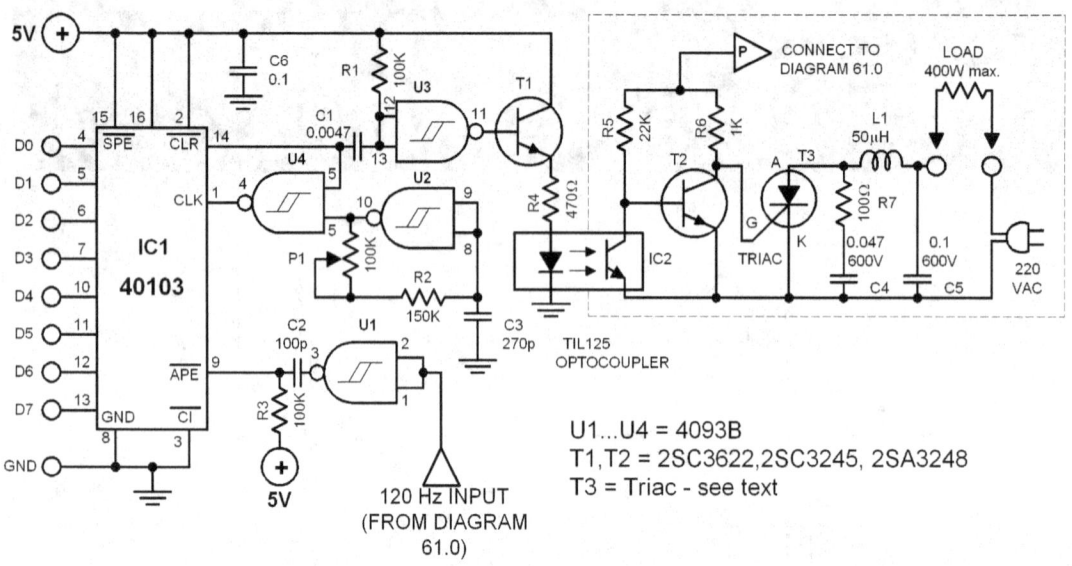

Diagram 61.1   Computer Controlled Power Regulation

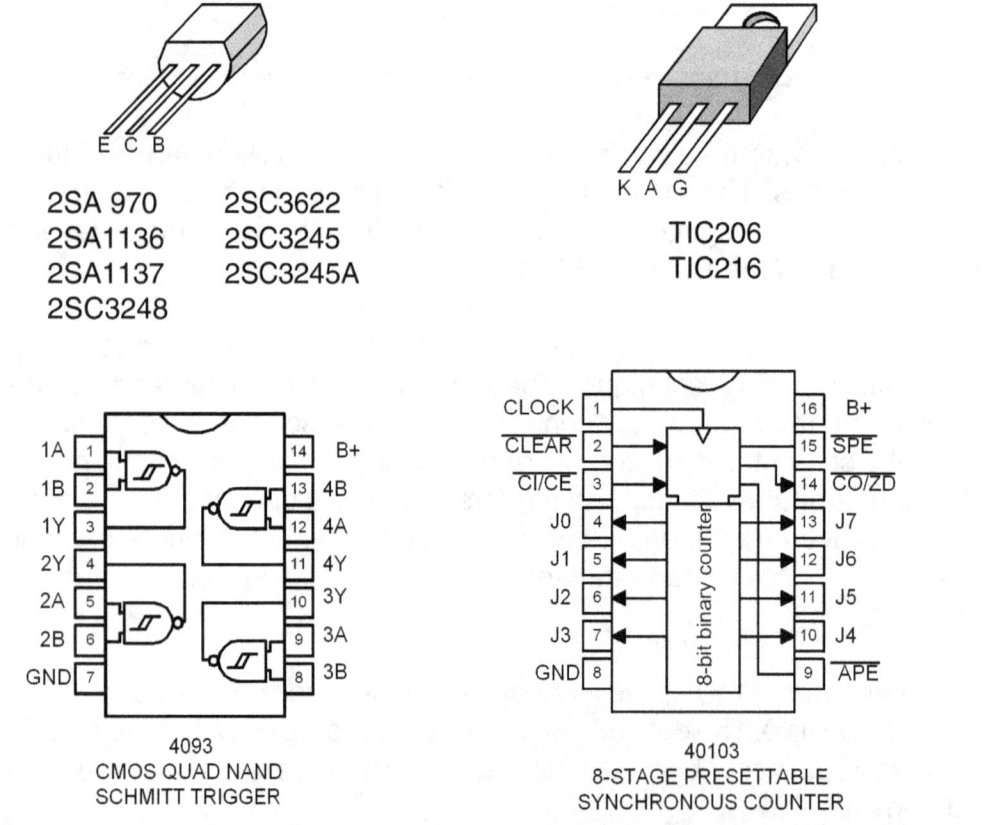

4093
CMOS QUAD NAND
SCHMITT TRIGGER

40103
8-STAGE PRESETTABLE
SYNCHRONOUS COUNTER

62 5A DELAYED POWER-ON

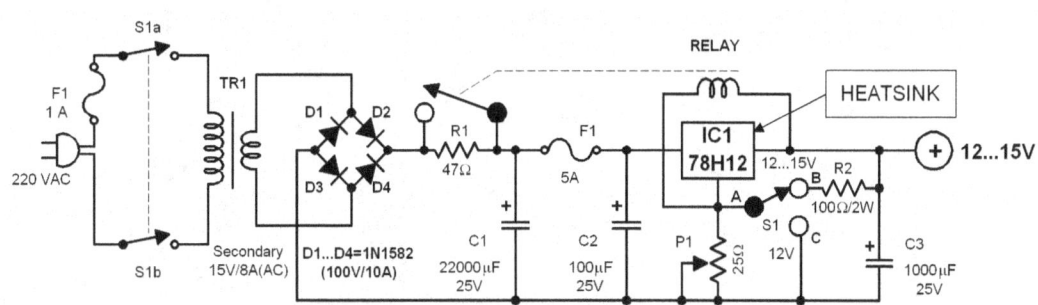

Diagram 62.0 5A Delayed Power-On

This power supply has a short power-on delay. This short delay has two advantages: First, the initial short high current present during switching a power supply with high value capacitors and high efficient transformers is avoided. Second, this short delay prevents the overvoltage fluctuations produced during switching from reaching the load circuits. This is very important for load circuits which are highly sensitive to voltage spikes.

The regulator can be operated in either fixed output mode or variable output mode. Switch S1 selects between the two modes. Position B is the fixed output mode and the power supply delivers a well regulated 12V. Position C is the variable mode and the output can be varied from 12 volts to 15 volts. The output voltage can be varied through P1. The maximum current output of T8H12 is 5A and the maximum short circuit current is around 7A. The IC must be heatsinked properly.

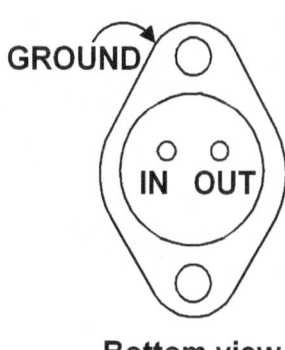

**Bottom view
78H12**

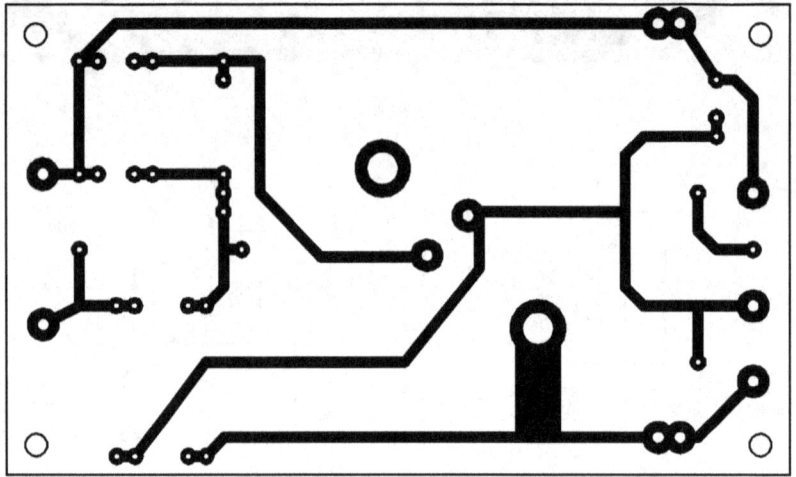

Figure 62.0   Printed Circuit Layout

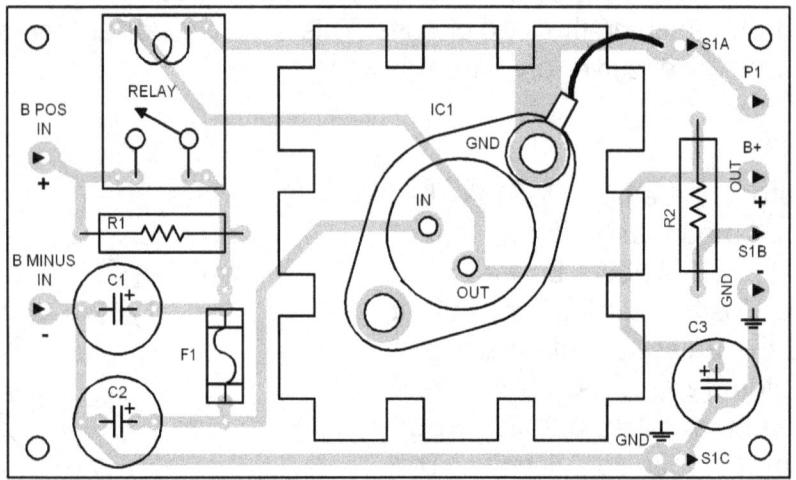

Figure 62.1   Parts Placement Layout

63 ROBUST 5V SUPPLY

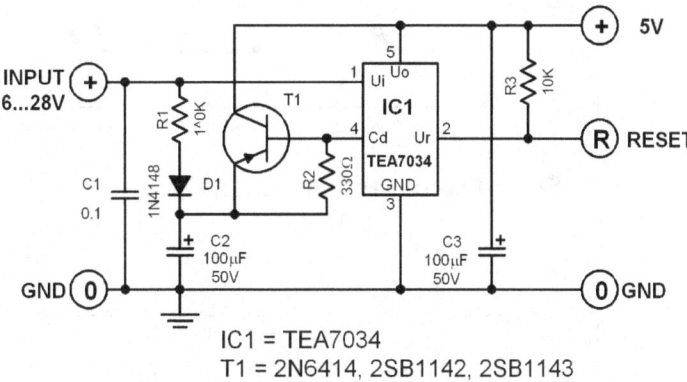

Diagram 63.0   Robust 5V Supply

This 5V supply circuit is very durable and insensitive to voltage fluctuations. It can block voltage spikes up to 80V. Very short current interference does not affect the function of the IC. This circuit is specially designed to support microprocessor circuits.

The maximum current output is 500 mA. The voltage drop between the input and the output voltages is around 0.6V. The output voltage is 5V+/-2.5%. The IC is also protected from thermal overloads.

Parts List:

Resistors:
R1,R3 = 10K
R2 = 330Ω
Capacitors:
C1 = 0.1/50V
C2,C3 = 100µF/50V

Diode:
D1 = 1N4148
IC:
IC1 = TEA7034

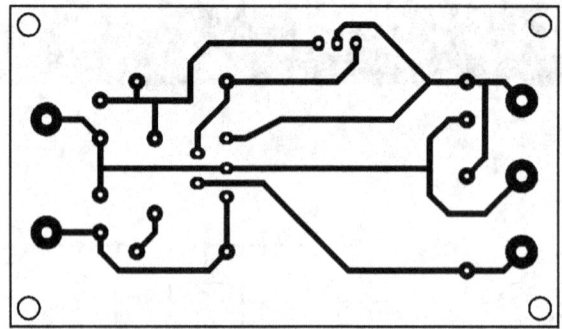

Figure 63.0 Printed Circuit Layout

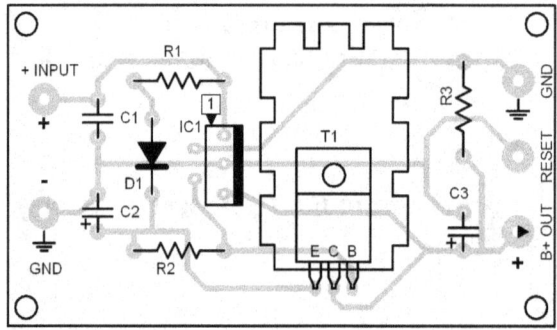

Figure 63.1 Parts Placement Layout

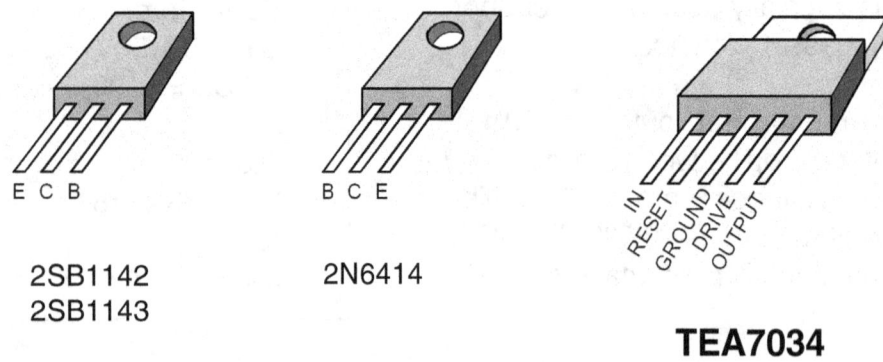

2SB1142
2SB1143

2N6414

TEA7034

64 AMPLIFIED REGULATOR

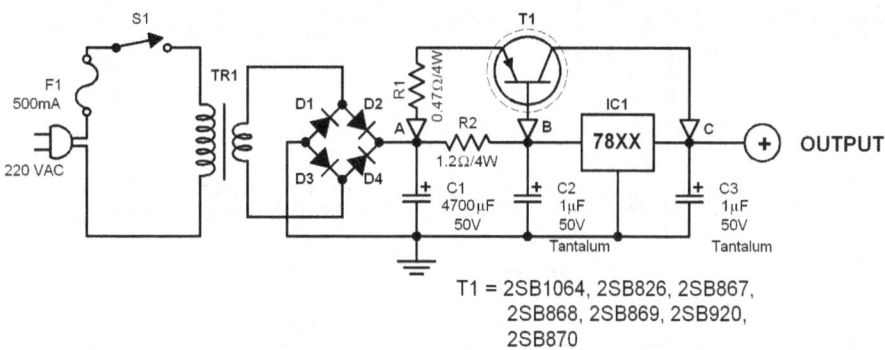

T1 = 2SB1064, 2SB826, 2SB867,
2SB868, 2SB869, 2SB920,
2SB870

Diagram 64.0   Amplified Regulator

Normally, 78xx voltage regulators can only deliver up to 1 ampere. This circuit can deliver currents up to 10 amperes by adding a current amplifier to the common 78xx series voltage regulators. The current amplifier can be a single or two power transistors in parallel depending on the needed maximum current output.

Table 64.0 shows the necessary component values for every maximum current output. The transistors must be heatsinked to minimize voltage drop caused by increasing temperature. When two transistors are used in parallel (additional $T1°$ transistor as shown in Diagram 64.1), a resistor $R1°$ must be added to each emitter terminal (see Diagram 64.1). The PCB layout for the higher current version is shown in figures 46.2 and 46.3. The transistor must be installed in a separate heatsink.

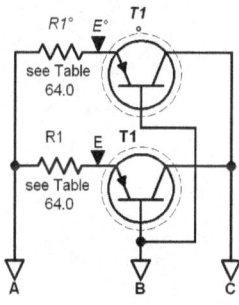

Diagram 64.1   Additional T1($T1°$)

**78xx
3-Terminal
voltage regulator**

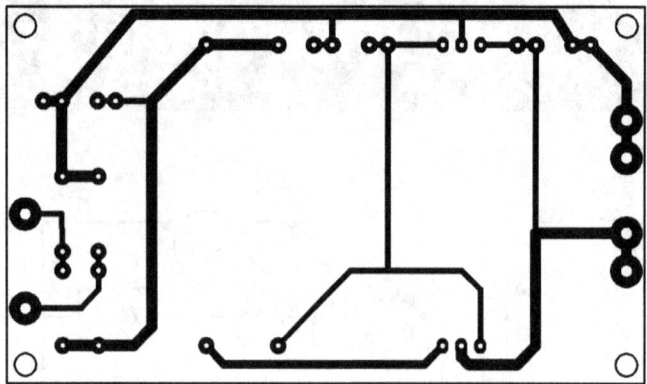

Figure 64.0 Printed Circuit Layout

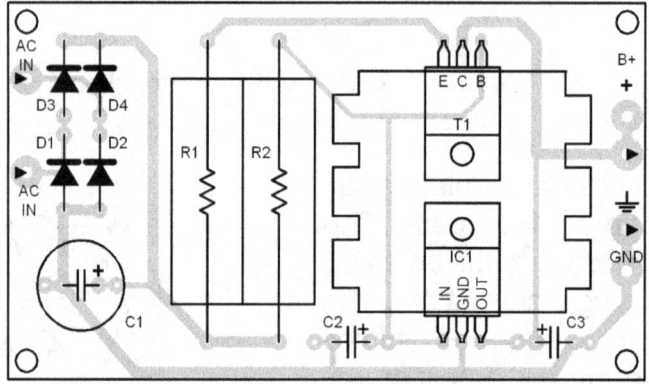

Figure 64.1 Parts Placement Layout

Maximum Current	D1...D4	C1 in µF	R1(*R1°*) in Ω/4W	R2 in Ω/4W	T1	*T1°*
2A	1N5401	4700 or 2 x 2200	0.47	1.2	2SB1064	
3 A	ECG5863	6800 or 3 x 2200	0.39	2.2	2SB638	
4 A	6...8A	10000 or 2 x 4700 or 4 x 2200	0.27	2.2	2SB638	
5 A	1N1582 8...10A	10000 or 2 x 4700 or 4 x 2200	0.22	2.2	2SB638	
7 A	10...14A	15000 or 3 x 4700	0.27	2.2	2SB638	2SB638
10 A	15...20A	22000 or 2 x 10000 or 4 x 4700	0.18	2.2	2SB638	2SB638
					See Diagram 64.0 for transistor equivalents.	

Table 64.0 Component values

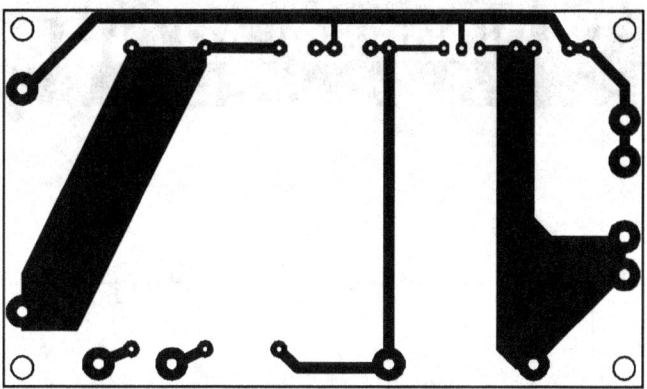

Figure 64.2 Printed Circuit Layout for
the higher current version

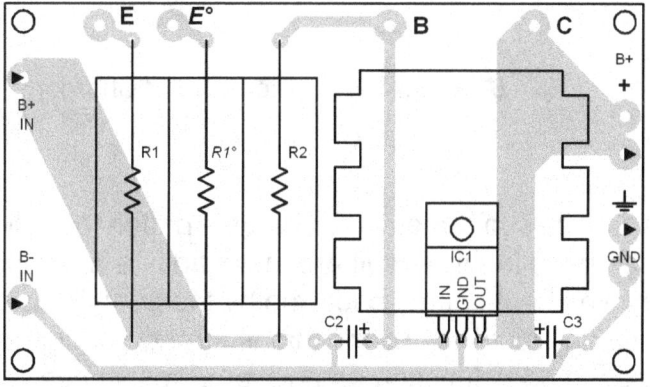

Figure 64.3 Parts Placement Layout for
the higher current version

2SB1064, 2SB826, 2SB867,
2SB868, 2SB869, 2SB920,
2SB870

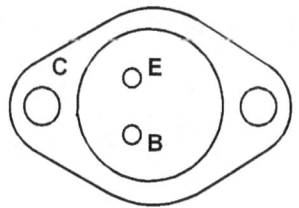

2SB638 2N6285
2SB639 2N6286

65 DIODE SOLAR CHARGER

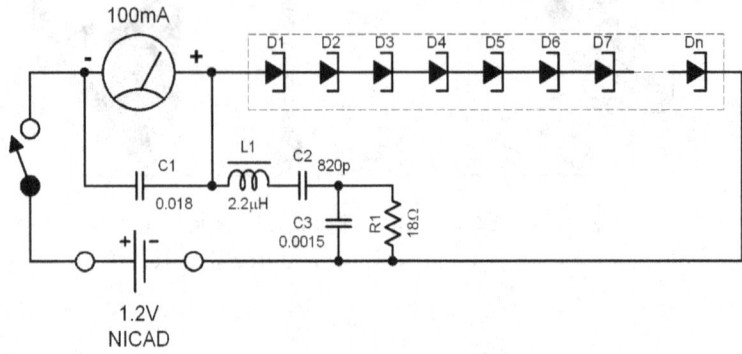

Diagram 65.0   Diode Solar Charger

You can directly tap the sun's energy to charge a battery by using tunnel diodes. The tunnel diodes used in this circuit are installed in a heatsink and left under the sun to warm. The heat energy captured by the heatsink is converted by the tunnel diodes into an electrical current that is strong enough to charge a nicad battery. The tunnel diodes function as a reverse resistance.

When the heatsink has a heat resistance of 3.5K/W and heated by the sunlight to around 26°C, the tunnel diodes can deliver a charging current of around 45 mA. Best results can be obtained by using a black heatsink. The heatsink must be nonmagnetic type (like aluminum) because any type of outer magnetic field (induced or permanent) can negatively affect the current carriers in a tunnel diode.

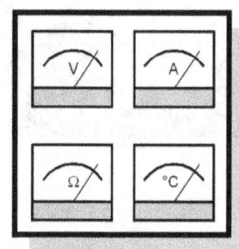

TESTERS & METERS

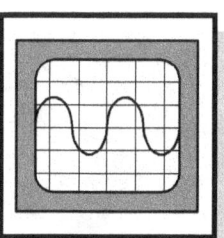

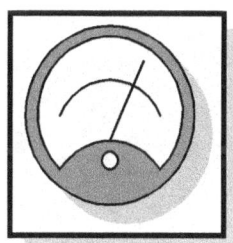

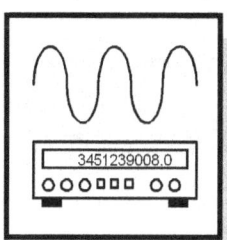

66   LINEAR OHMMETER

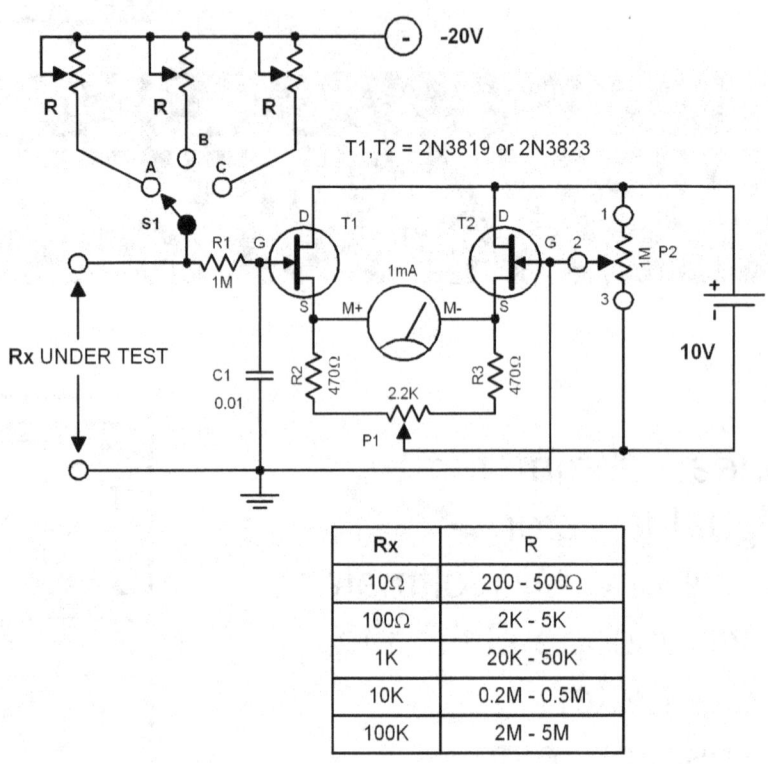

Rx	R
10Ω	200 - 500Ω
100Ω	2K - 5K
1K	20K - 50K
10K	0.2M - 0.5M
100K	2M - 5M

Diagram 66.0   Linear Ohmmeter

Ohmeters are always used in electronic work to measure resistances. Common multimeters, however, often do not have a linear scale. The circuit featured here offers the capability to measure resistances from 10 ohms up to 100 Megaohms in 5 ranges and displays the result linearly.

The meter is connected between two FET transistors in a bridge circuit configuration. The potentiometer 1M sets the balance of the circuit. Zero adjust is done through the 2.2K potentiometer. The trimmers R are used for calibrating each range.

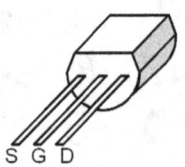

2N3819

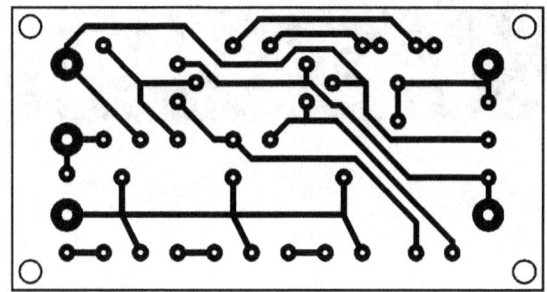

Figure 66.0 Printed Circuit Layout

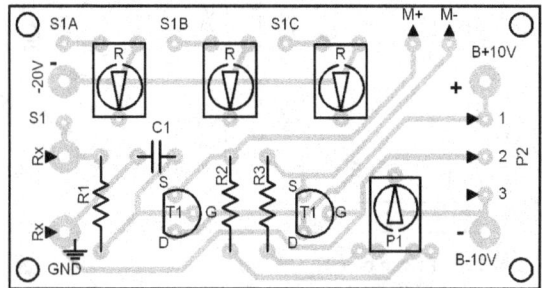

Figure 66.1 Parts Placement Layout

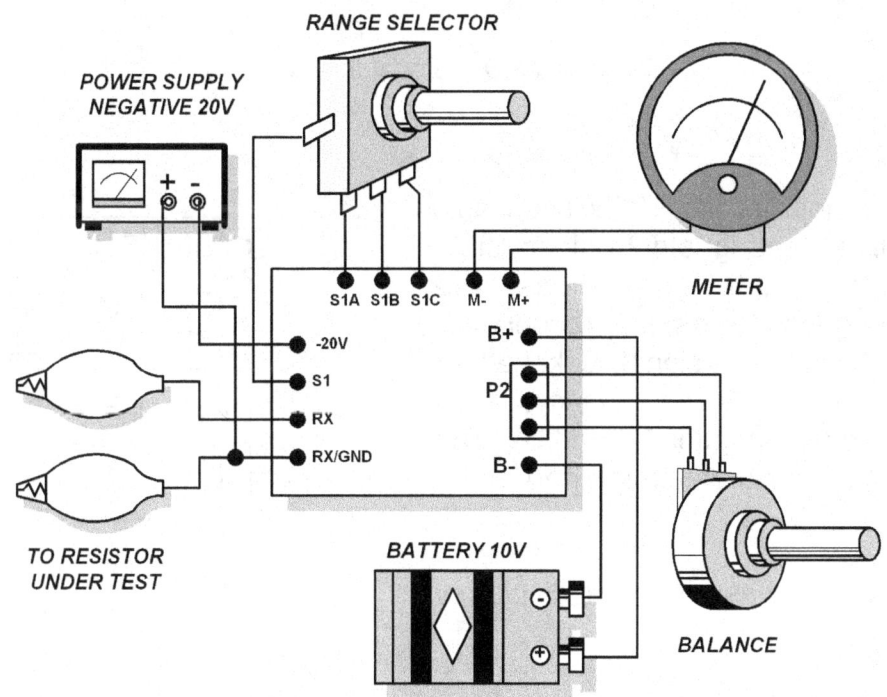

Figure 66.2 External Wirings

67 SIGNAL INJECTOR

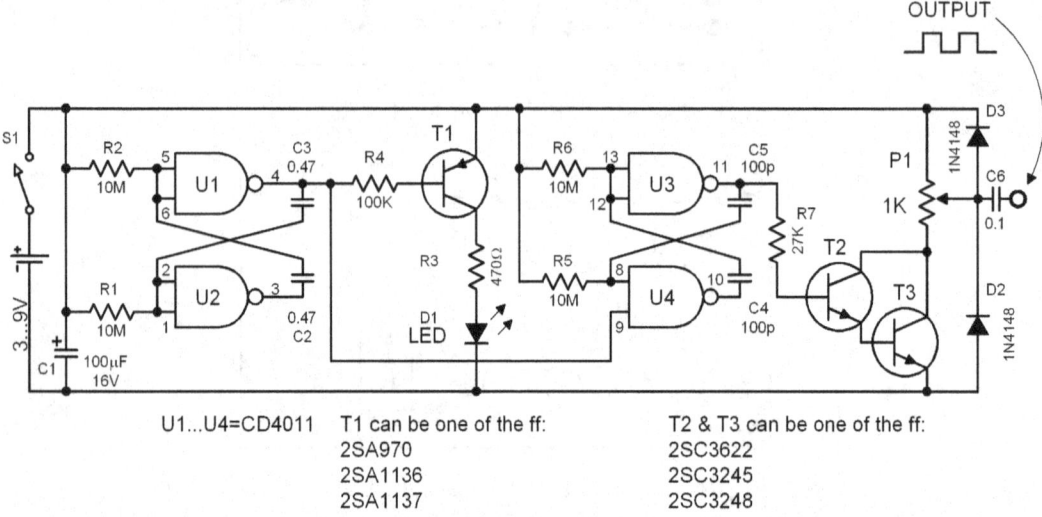

Diagram 67.0   Signal Injector

U1...U4=CD4011   T1 can be one of the ff: T2 & T3 can be one of the ff:
2SA970   2SC3622
2SA1136   2SC3245
2SA1137   2SC3248

This signal injector circuit is indispensable in troubleshooting electronic circuits specially in audio applications. The oscillator generates a signal frequency of 1 kHz. This main signal is keyed or modulated by a low frequency signal generated by the gates U1 and U2. The lower frequency is determined by C2 and C3.

LED D3 serves as an optical display of the keying.

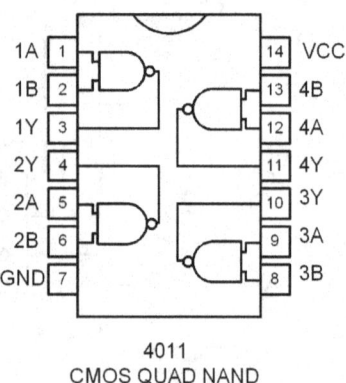

4011
CMOS QUAD NAND

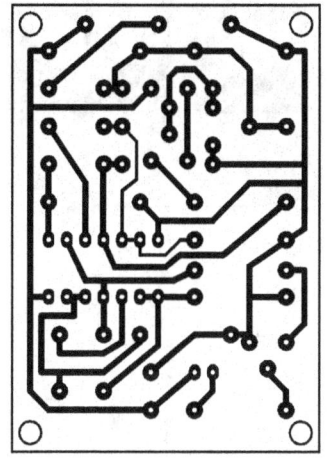

Figure 67.0 Printed Circuit Layout

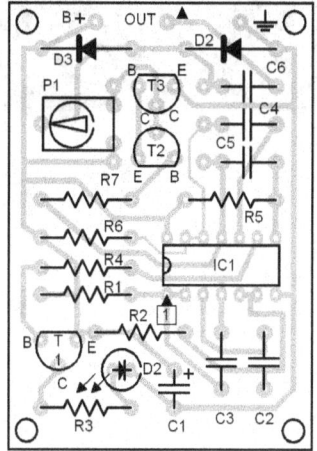

Figure 67.1 Parts Placement Layout

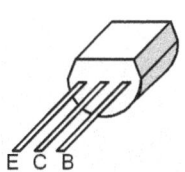

2SA 970	2SC3622
2SA1136	2SC3245
2SA1137	2SC3245A
2SC3248	

Parts List:

R1,R2,R5,R6 = 10M	C4,C5 = 100p/50V ceramic
R3 = 470Ω	C6 = 0.1/50V ceramic
R4 = 100K	D1 = LED
R7 = 27K	D2,D3 = 1N4148
P1 = 1K trimmer	T1 = 2SA970(2SA1136)
C1 = 0.1/50V	T2 = 2SC3622(2SC3245)
C1 = 100µF/16V	IC1 = 4011
C2,C3 = 0.47/50V ceramic	

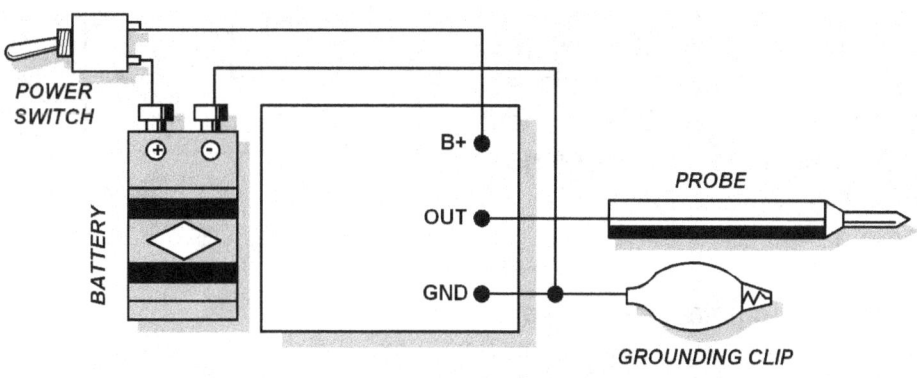

Figure 67.2 External Wirings

68 WIDEBAND MILLIVOLTMETER

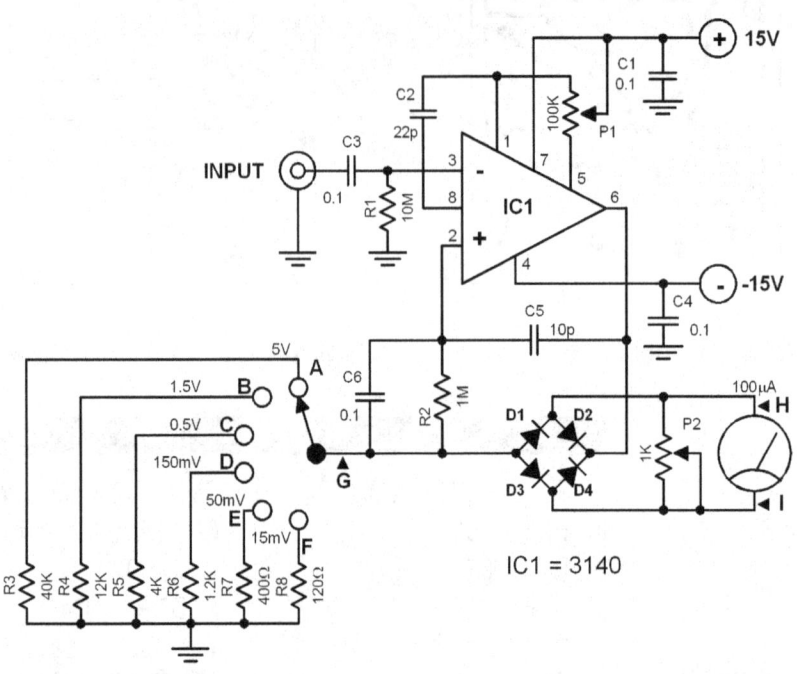

Diagram 68.0   Wideband Millivoltmeter

This millivoltmeter functions very well in measuring AC frequencies between 100 Hz and 500 kHz. The input impedance is very high in all ranges - 10 Megaohms- due to the application of an OPAMP with a MOSFET input. The sensitivity in the lowest range is 15 mV, strong enough to drive the 100 µA meter to full deflection.

P2 is adjusted to obtain the zero point while the probes are short circuited. The calibration of the measuring range is done through P1.

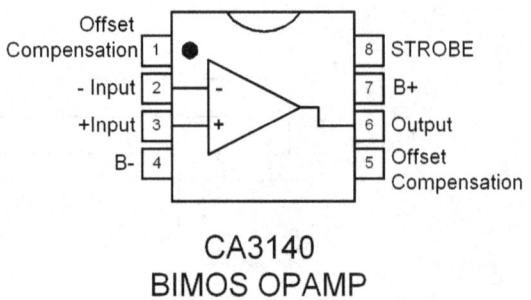

CA3140
BIMOS OPAMP

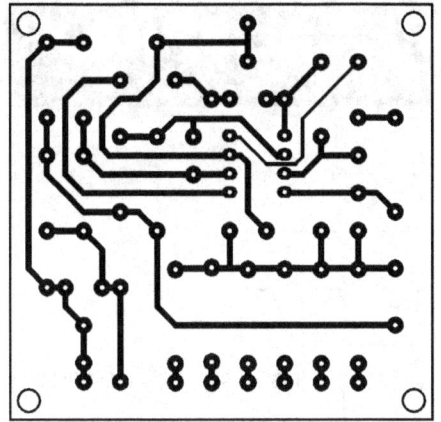

Figure 68.0 Printed Circuit Layout

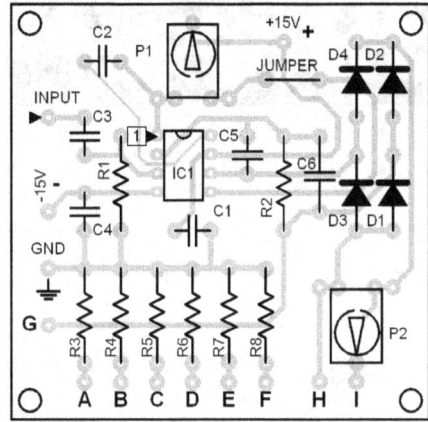

Figure 68.1 Parts Placement Layout

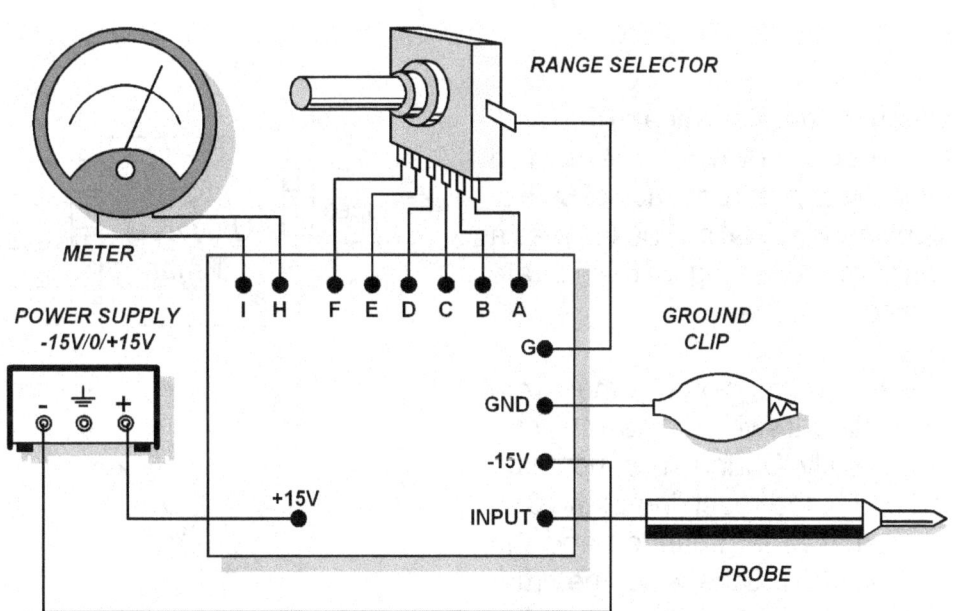

Figure 68.2 External Wirings

69 OSCILLOSCOPE CALIBRATOR

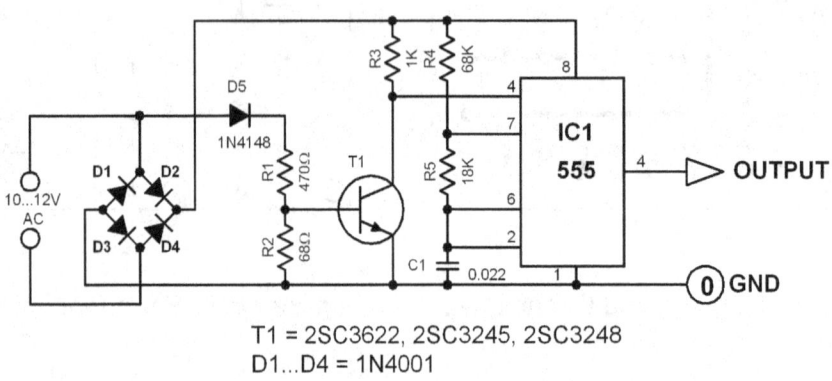

T1 = 2SC3622, 2SC3245, 2SC3248
D1...D4 = 1N4001

Diagram 69.0   Oscilloscope Calibrator

This circuit generates a chain of pulses. The pulse chain is repeated periodically with a frequency of 60 Hz. This pulsating signal can be used as a standard reference signal to calibrate oscilloscope probes.

The 60 Hz control signal comes from the power line. It switches the transistor T1 periodically. During the conduction period of T1, the reset line of the 555 timer IC (pin 3) is grounded and the timer is blocked. Otherwise, the timer is released and it functions as stable multivibrator producing the series of short pulses.

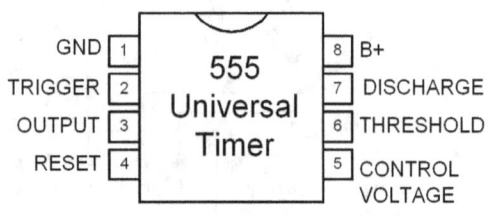

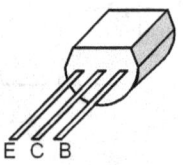

2SC3622
2SC3245
2SC3245A
2SC3248

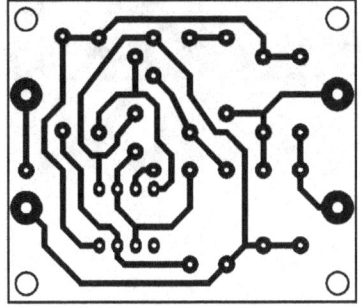

Figure 69.0 Printed Circuit Layout

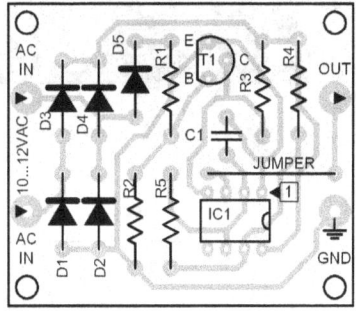

Figure 69.1 Parts Placement Layout

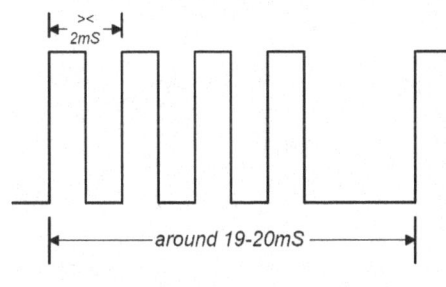

Output signal of the
oscilloscope calibrator

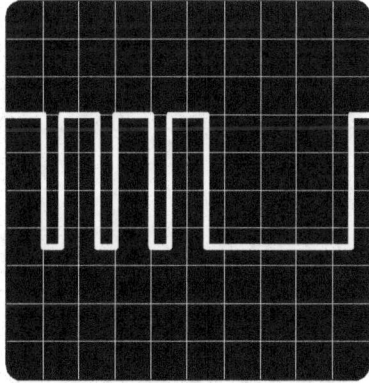

Output signal as seen on
the oscilloscope screen

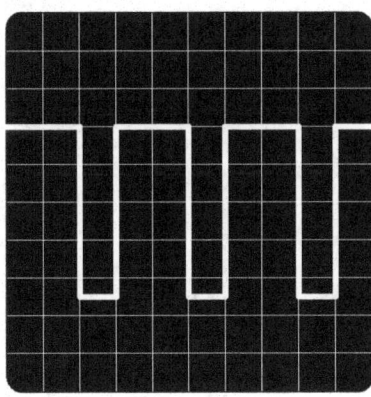

Waveforn when the probe
is calibrated

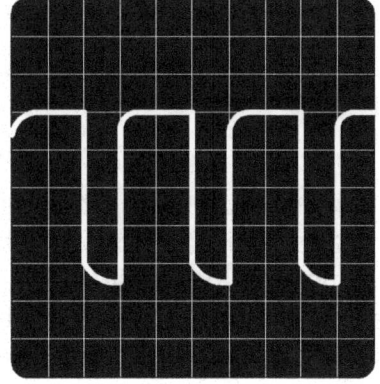

Waveform when the probe
is not calibrated

70 L & C METER

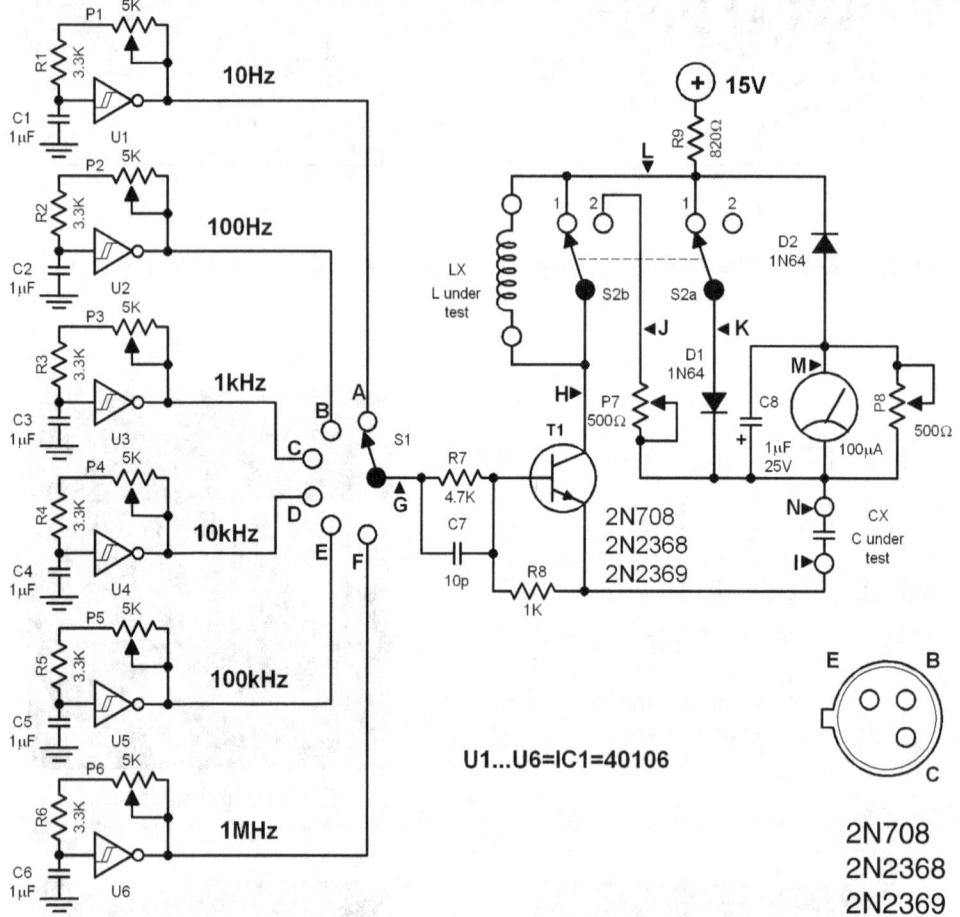

Diagram 70.0 L & C Meter

U1...U6=IC1=40106

2N708
2N2368
2N2369

2N708
2N2368
2N2369

This circuit measures the inductivity of a coil and the capacitance of a capacitor using the simplest method. Inductivity is measured (S2 in position 1) through the technique of periodically pulsing current in the unknown coil. When S2 is switched to position 2, the capacitance is measured.

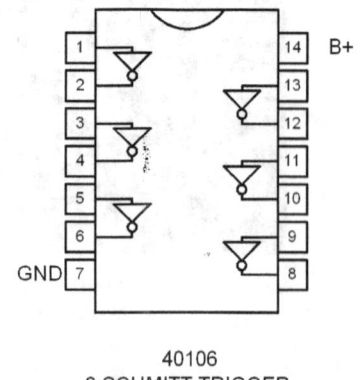

40106
6 SCHMITT TRIGGER

S1 selects the measuring range. Calibration: With the help of either an oscilloscope or a frequency counter, set the potentiometers P1...P6 so that the oscillators generate their correct frequencies. Then connect a capacitor of a known value like 100 pf in the Cx terminal and set P7 until the meters shows the value of the capacitor. Switch S2 to position 2 and connect an inductance of a known value (like 100µH) in the Lx terminals and set P8 until the meter shows the value of the coil.

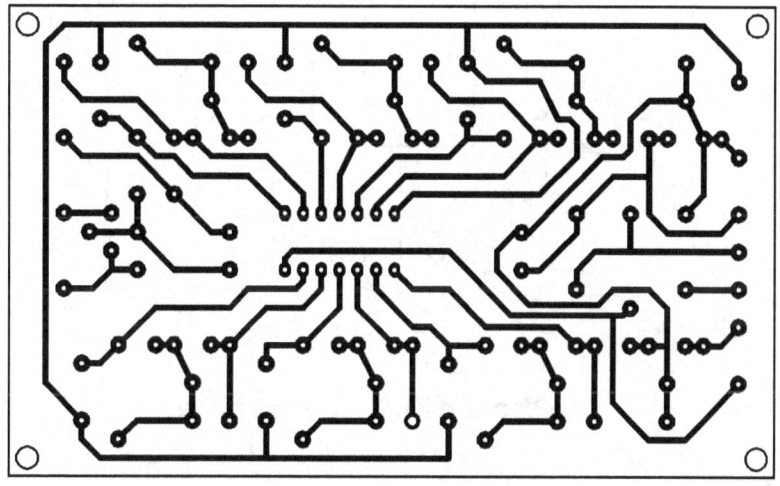

Figure 70.0   Printed Circuit Layout

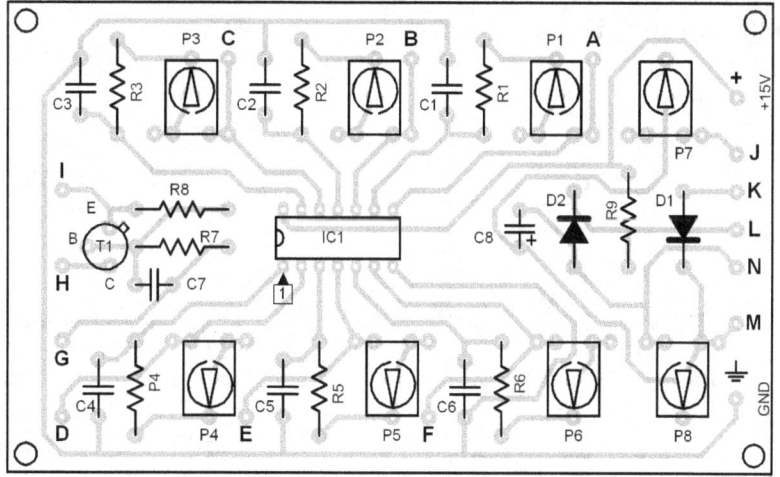

Figure 70.1   Parts Placement Layout

Page 137

71 TRANSISTOR TESTER

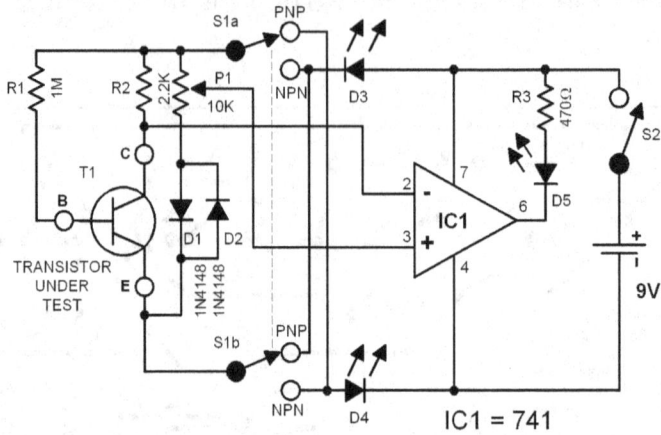

Diagram 71.0   Transistor Tester

The current amplification of a transistor is called beta. You can measure it by using this circuit.   The measuring capability of the tester is independent from its supply voltage. The current flows to the base of the Transistor-under-Test (TUT) through resistor R1.

The base current can be found with:

$$Ibase = \frac{Vxy - Vbe}{R1}$$

The voltage drop at the collector's resistor can be found with:

$$Vr2 = Ic(R2) = Beta(Ib(R1))$$

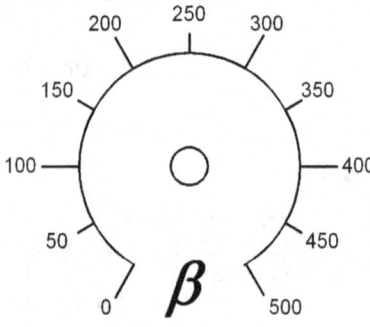

Scale template for potentiometer P1

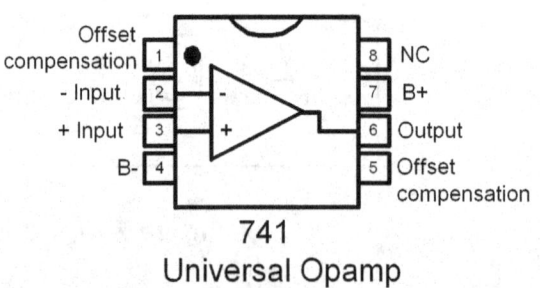

741
Universal Opamp

The potentiometer P1 shifts the voltage drop of D1 (D2 for PNP) to the base-emitter voltage of the TUT. The setting of potentiometer P1 represents the measured current amplification of the TUT. The OPAMP is wired as a comparator, and compares the voltage drop at R2 with the sampled voltage at P1. The pot P1 is adjusted until it reaches the position between the "on" and "off" states of the LED D5.

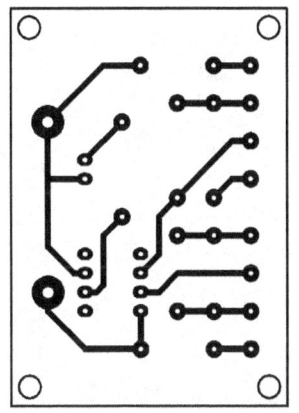

Figure 71.0   Printed Circuit Layout

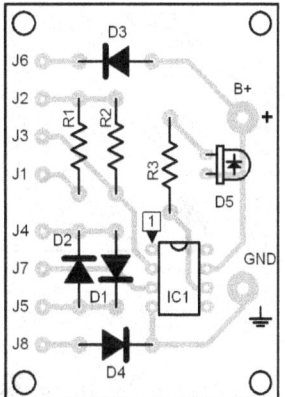

Figure 71.1   Parts Placement

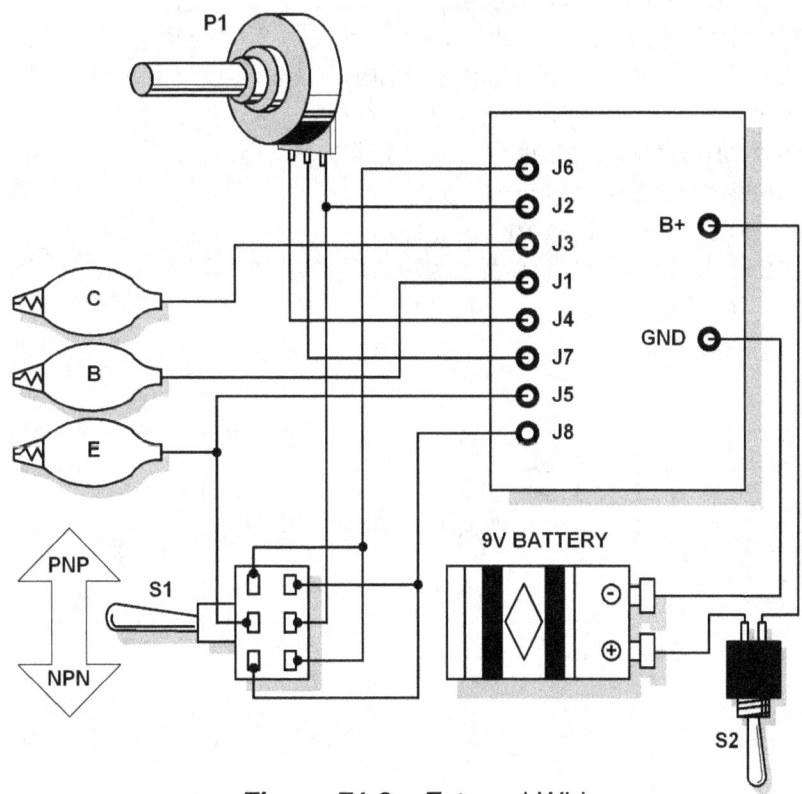

Figure 71.2   External Wirings

72 ADAPTIVE LOGIC PROBE

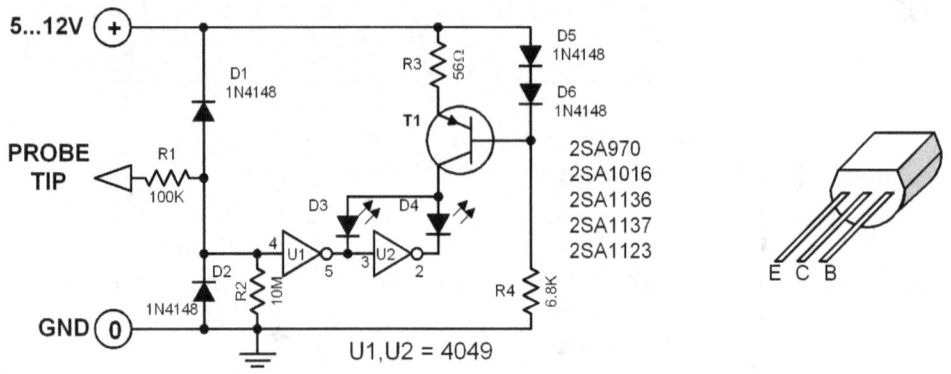

Diagram 72.0 Adaptive Logic Probe

This circuit is very simple to construct. Its function is also very simple. However, simple it may be, it is a helpful tool in troubleshooting digital circuits. The red LED D3 lights when the tested logic is "1" or "high" and the green LED D4 lights when the logic is "0" or "low". The supply voltage of the logic probe comes from the circuit being tested. With this technique, the logic probe adapts itself automatically to the voltage levels of the circuits being tested.

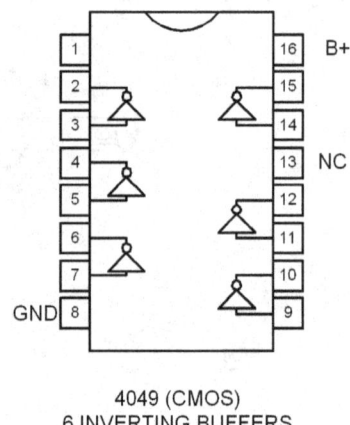

4049 (CMOS)
6 INVERTING BUFFERS

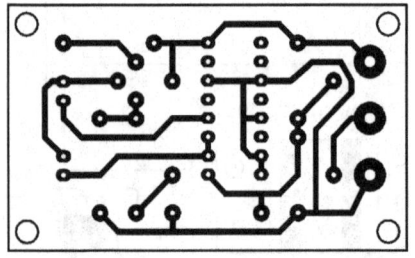

Figure 72.0 Printed Circuit Layout

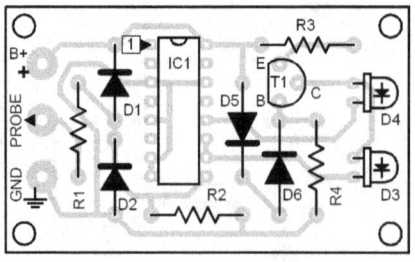

Figure 72.1 Parts Placement

73 OSCILLOSCOPE MULTIPLEXER

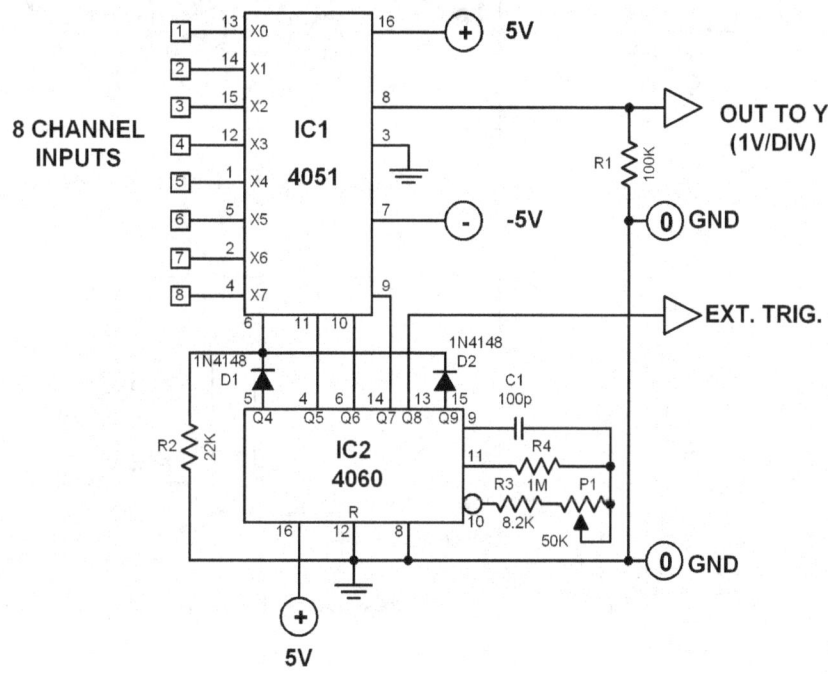

8 CHANNEL INPUTS

Diagram 73.0 Oscilloscope Multiplexer

You have an oscilloscope. Well and good. What if it is necessary to sample several voltage levels all at once? No need to worry, just use the oscilloscope multiplexer featured here. This circuit enables your single channel oscilloscope to simultaneously monitor eight independent DC voltage levels. It uses a 4051 IC - an 8-channel analog multiplexer which is driven by a binary counter 4060. This counter in turn, is driven by an internal clock generator with a frequency of 60 kHz. The frequency is adjustable through P1.

The different voltages to be monitored are connected to the multiplexer inputs X0 to X7. The time base of the oscilloscope must be set to 0.5mS per division. The triggering must be set to positive. The input voltage can be set between -4V and +4V. The input sensitivity of the circuit is around 1V/div.

Calibration: Adjust P1 until all 8 channels are fully accomodated in the oscilloscope screen.

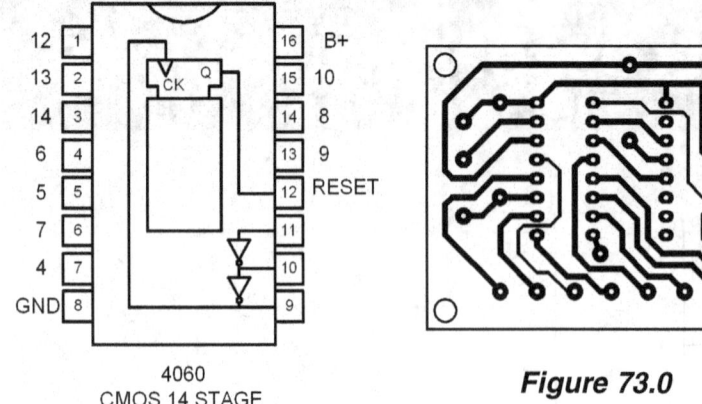

4060
CMOS 14 STAGE
BINARY COUNTER

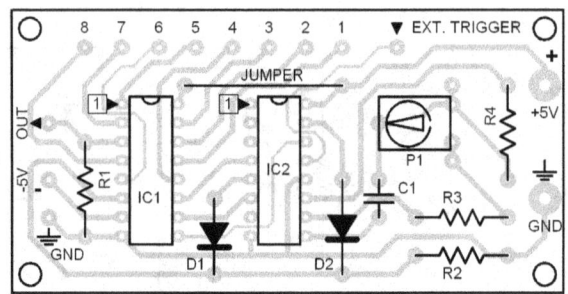

Figure 73.0 Printed Circuit Layout

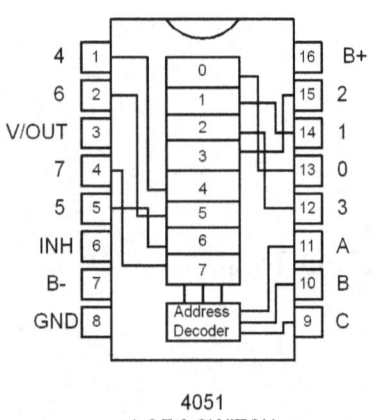

4051
1 OF 8 SWITCH

Figure 73.1 Parts Placement Layout

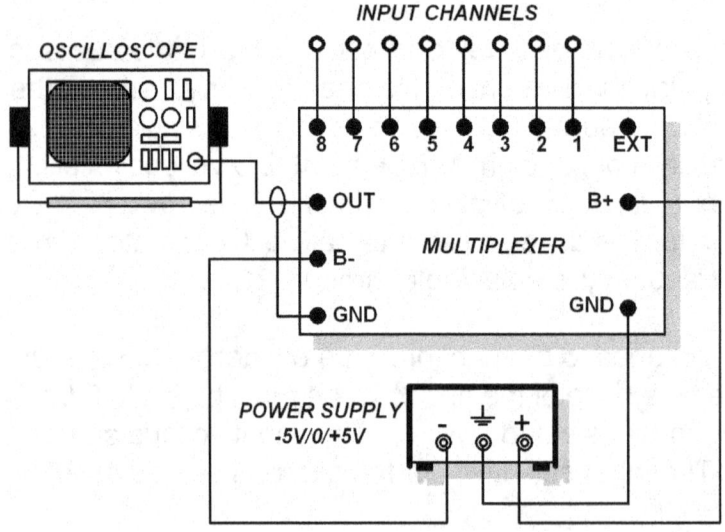

Figure 73.2 External Wirings

74 SWEEP GENERATOR

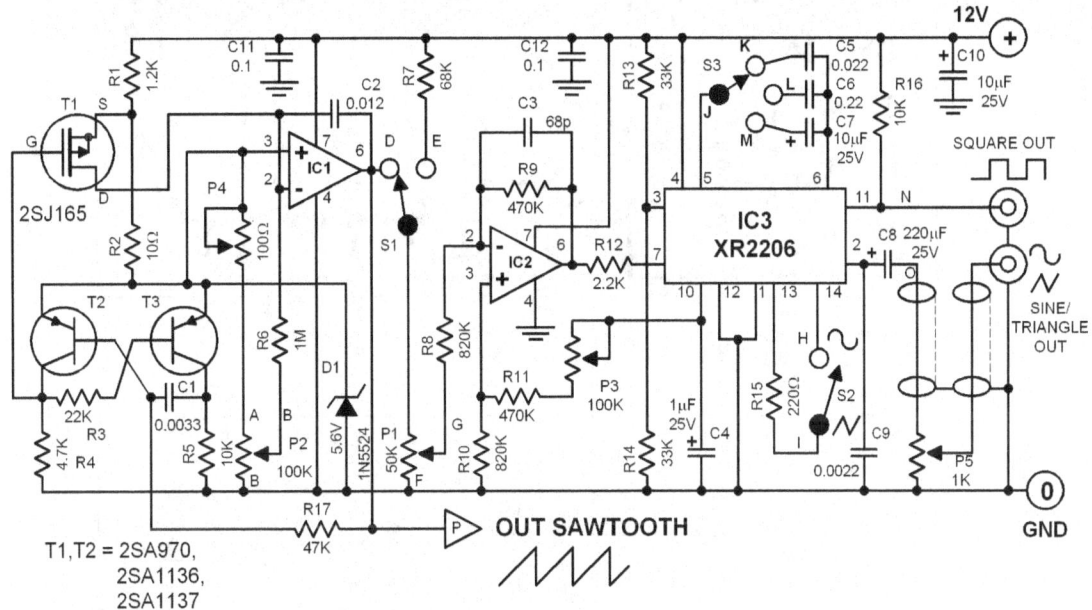

Diagram 74.0 Sweep Generator

A sweep generator is a highly invaluable instrument in audio applications. It is usually used to measure the frequency bandwidth of an amplifier in a fast way. The generator circuit featured here uses an XR2206 IC with 3 selectable capacitors to set the frequency range. The output level is adjustable through P5. The left part of the circuit is a sawtooth generator and a buffer stage for the 2206. The sawtooth level is adjustable through P1. P3 is used to calibrate the circuit. The sweep generator delivers squarewave, sinewave, and sawtooth signals.

The potentiometers control the following functions:

P1 = sawtooth output level adjust
P2 = time constant adjust (0.01 sec. to 10 sec.)
P3 = frequency scale adjust
P4 = time constant maximum adjust
P5 = output level

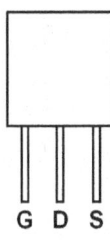

2SJ165

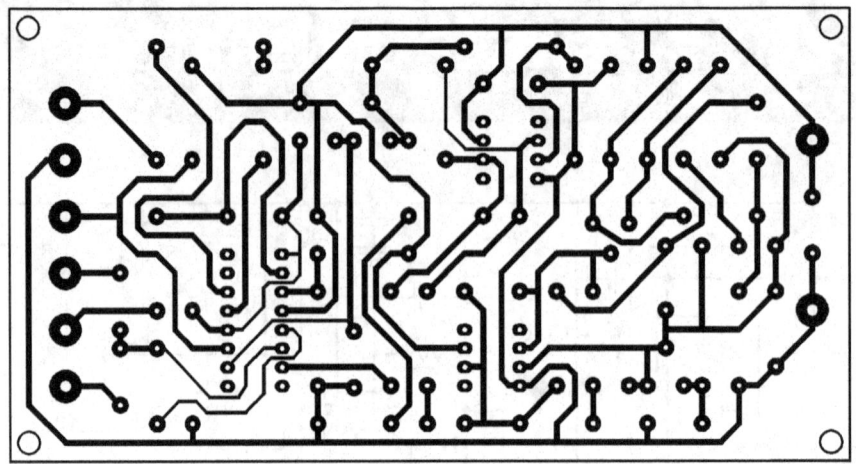

Figure 74.0 Printed Circuit Layout

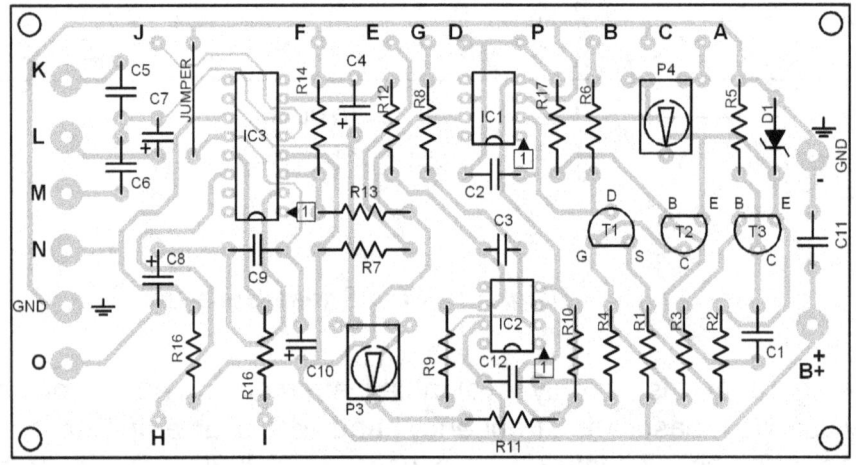

Figure 74.1 Parts Placement Layout

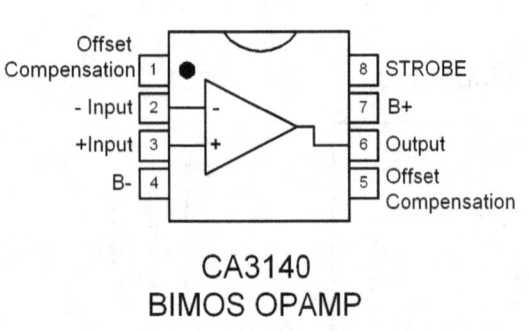

CA3140
BIMOS OPAMP

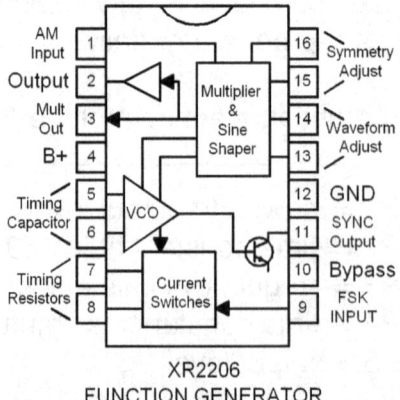

XR2206
FUNCTION GENERATOR

75 DIGITAL PANEL METER

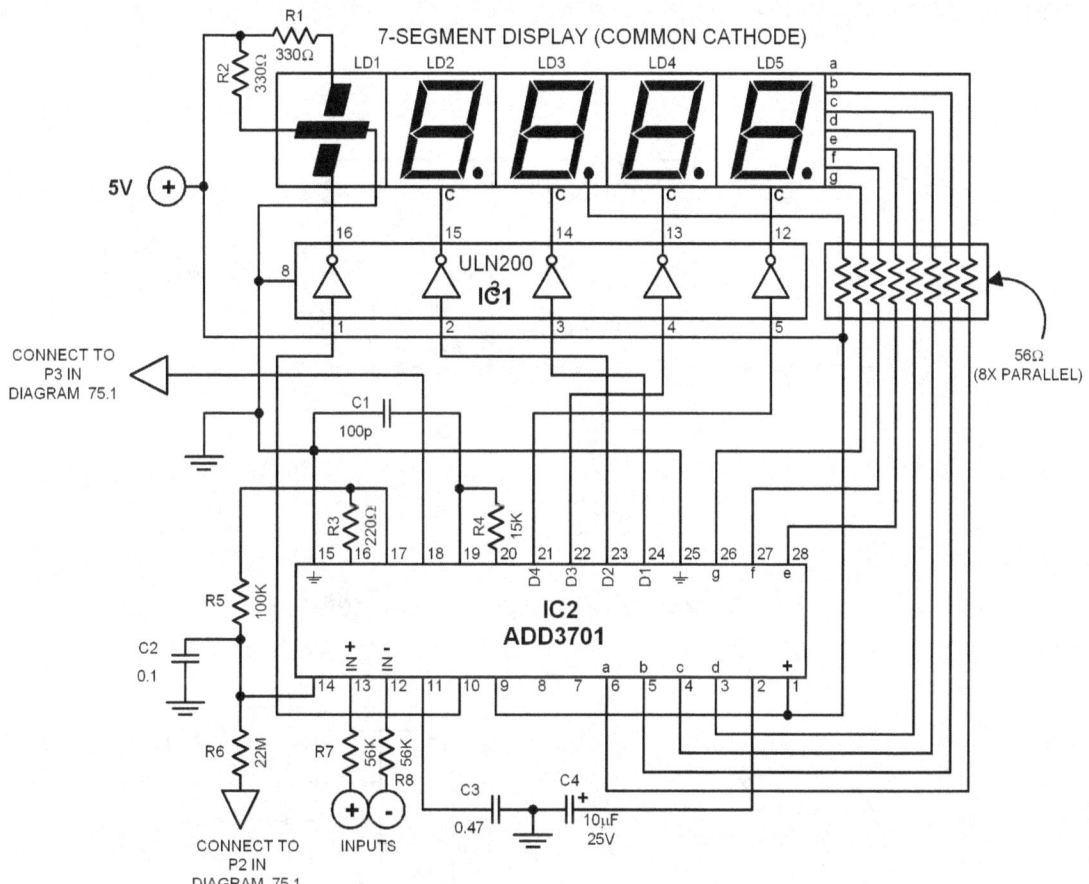

Diagram 75.0   Digital Panel Meter

This panel meter has a 3 and 3/4 display capacity. Its display range is from -3999 up to +3999. This very wide display range is made possible through a special IC - the ADD3701. This IC has an integrated A/D converter and a display multiplexer. The LED display has a common cathode connection. The frequency of the integrated clock oscillator is determined by the RC combination R7 and C6. The clock frequency determines in turn the sampling rate. The clock frequency of the given circuit is around 400 kHz. This generates a sampling rate of 3 samples per second.

Calibration: Short the input terminals and adjust P2 until the LEDs display 0000. Remove the short at the input. Then connect a 1.9V supply at the input and adjust P3 until the display shows 3.800. Note that the displayed value is twice that of the measured value. This must be taken into account in designing the necessary voltage divider circuits. The supply voltage for the circuit can be between 8 and 12 volts.

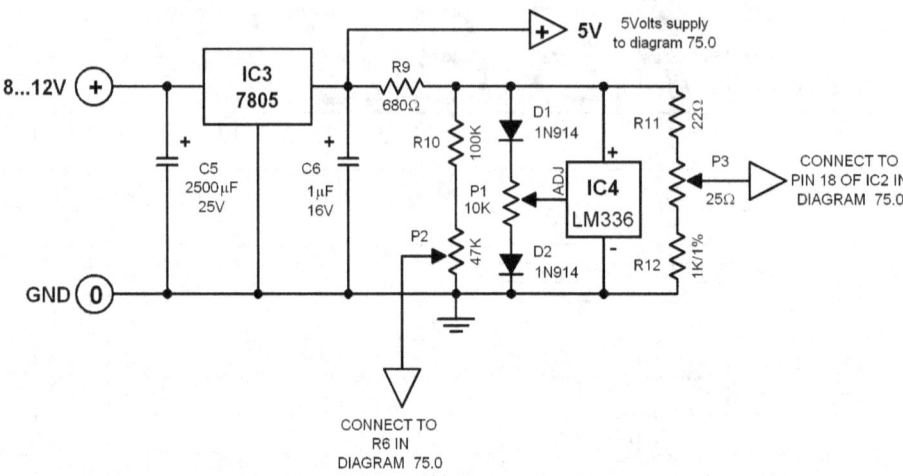

Diagram 75.0 Digital Panel Meter

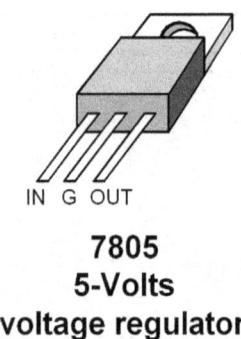

**7805
5-Volts
voltage regulator**

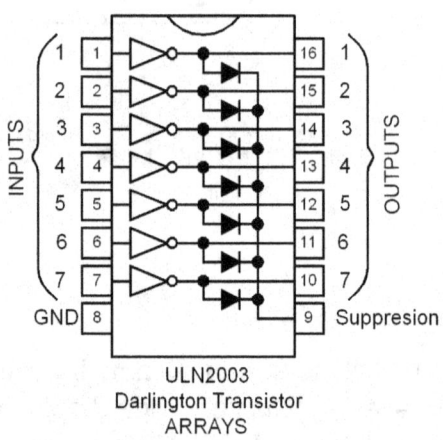

**ULN2003
Darlington Transistor
ARRAYS**

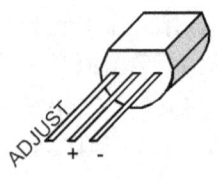

**LM336
Z-PACKAGE
TO-92 PLASTIC**

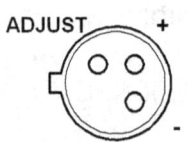

**LM336
H-Package
TO-46 Metal Can**

76 VOLTAGE MONITOR

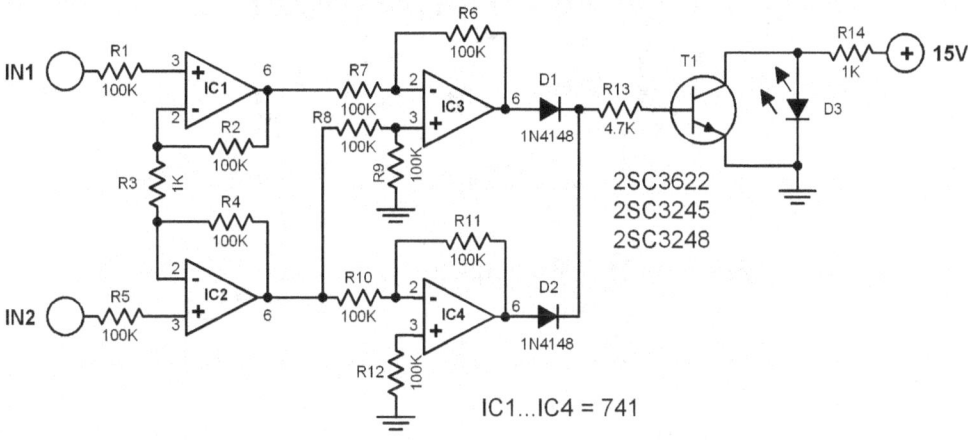

Diagram 76.0 Voltage Monitor

This circuit monitors a certain voltage level and gives out an optical signal when the level goes beyond the preset tolerances. The monitoring is done by comparing the monitored voltage with a very stable reference. When the difference between the two voltage levels is higher than 10mV, the LED turns off signalling that something is wrong with the monitored voltage.

The voltage level being monitored is connected to terminal 1 and the reference voltage is connected to terminal 2. The circuit's supply voltage is not critical.

Parts List:
R1,R2,R4,R5 = 100K
R6,R7,R8,R9 = 100K
R10,R11,R12,R13 = 100K
R3,R15 = 1K
R14 = 4.7K
D1,D2 = 1N4148
D3 = LED
T1 = 2SC3622(2SC3245)
(2SC3248)
IC1,IC2,IC3,IC4 = 741

A symmetrical supply (+-15V) can be used as well as a single +15V supply. However it must be taken into account that the input voltages must always be 1.5V lower than the supply voltage.

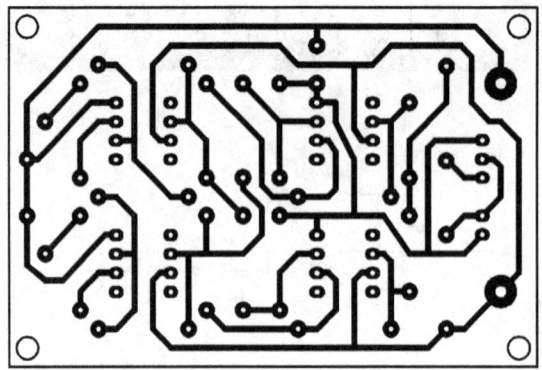

Figure 76.0   Printed Circuit Layout

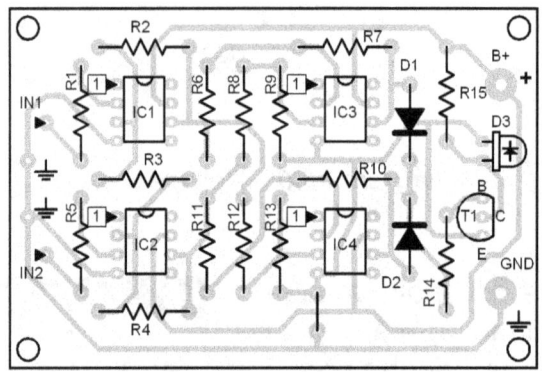

Figure 76.1 Parts Placement Layout

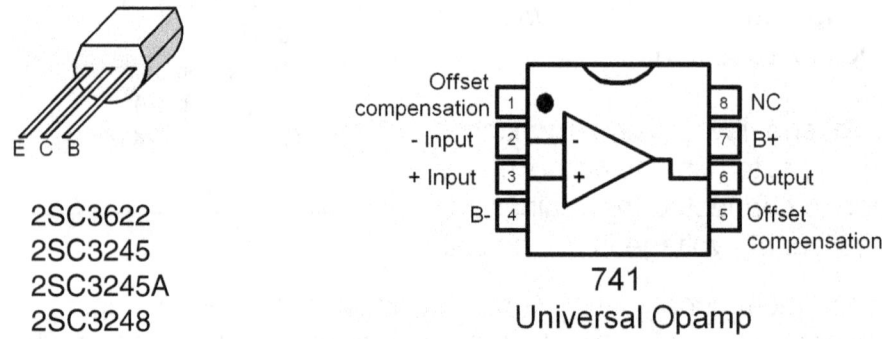

2SC3622
2SC3245
2SC3245A
2SC3248

741
Universal Opamp

77 555 IC TESTER

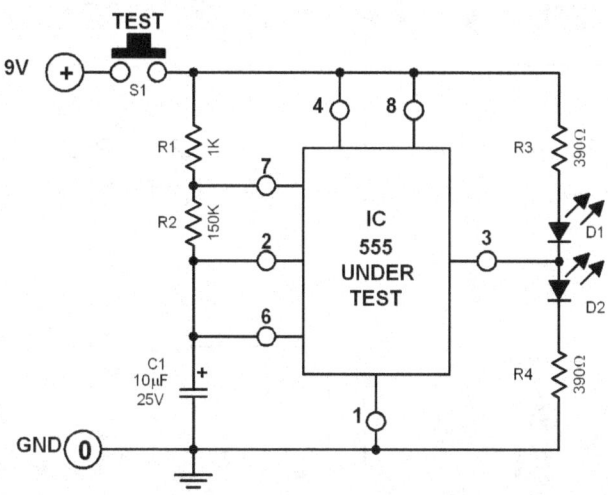

Diagram 77.0   555 IC Tester

Although 555 ICs belong to the bipolar family of ICs and relatively immune to mishandling, they sometimes go crazy. This circuit tests the IC whether it functions or not.

When the IC is functioning, the two LEDs blink one after the other. Otherwise, the IC is defective. The blink frequency (f) can be found with the following formula:

$$f(hz) = \frac{1.44}{(R1+(2R2))C1}$$

where   C= µF
        R= kiloohm

The component values given in the circuit causes the IC to generate a blink frequency of around 0.5 Hz.

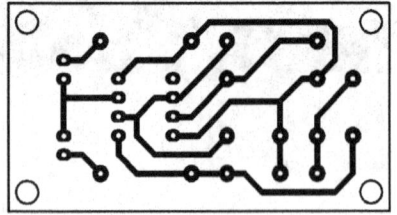

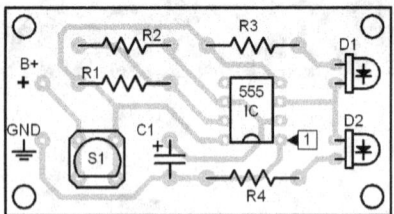

Figure 77.0 Printed Circuit Layout of 555 IC Timer

Figure 77.1 Parts Placement Layout of 555 IC Timer

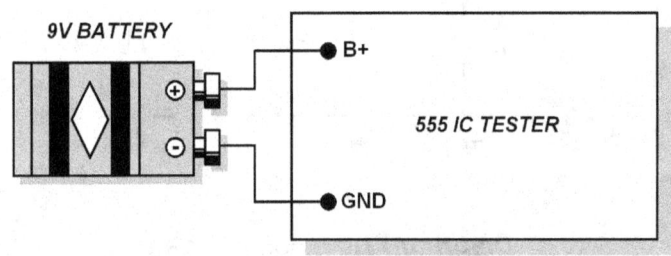

Figure 77.2 External Wiring of 555 IC Timer

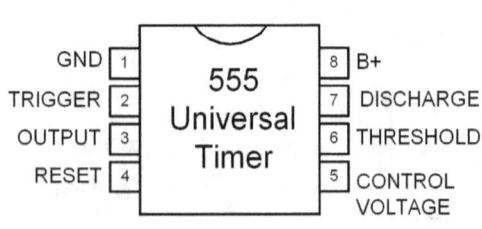

Parts List:

R1 = 1K
R2 = 150K
R3,R4 = 390Ω
R7,R8 = 39K
C1 = 10μF/25V
D1,D2 = LED
IC1 = 555 Timer IC

78 AUDIO SCOPE

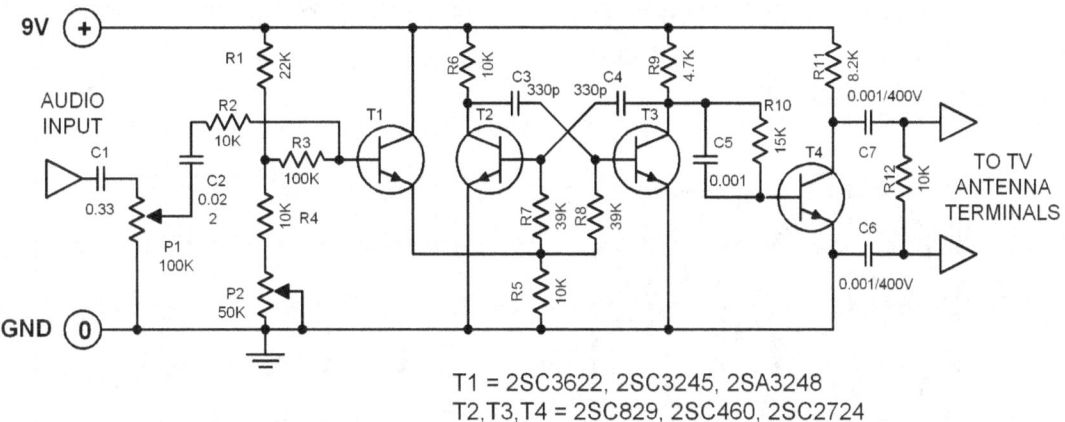

T1 = 2SC3622, 2SC3245, 2SA3248
T2,T3,T4 = 2SC829, 2SC460, 2SC2724

Diagram 78.0 Audio Scope

This circuit enables audio signals to be seen and monitored through an ordinary television set. When using this circuit, one only needs to connect it to the VHF antenna terminals of the TV set. This can be used as an audio-visual signal tracer in troubleshooting jobs. The circuit generates black vertical stripes on the TV screen. These stripes move in rhythm with the audio signal.

The heart of the circuit is an astable multivibrator which generates a frequency of around 16 kHz. This frequency is adjustable through P2 and is set to synchronize with the horizontal deflection frequency of the television set.

Parts List:

R1 = 22K
R2,R4,R5,R6,R12 = 10K
R3 = 100K
R7,R8 = 39K
R9 = 4.7K
R10 = 15K
R11 = 8.2K
P1 = 100K trimmer
P2 = 50K trimmer
C1 = 0.33/50V
C2 = 0.022/50V
C3,C4 = 330p/50V
C5,C6,C7 = 0.001/400V
T1 = 2SC3622(2SC3245)
    (2SC3248)
T2,T3,T4 = 2SC829(2SC460)
    (2SC2724)

In calibrating the circuit, set P1 to extreme left (thereby shorting input to ground) and set P2 until stable vertical stripes appear on the screen. The vertical stripes will move in horizontal direction according to the strength of the input signal. A 9V battery can be used as a power supply.

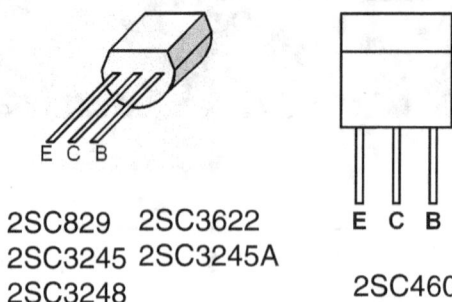

2SC829 2SC3622
2SC3245 2SC3245A
2SC3248

2SC460

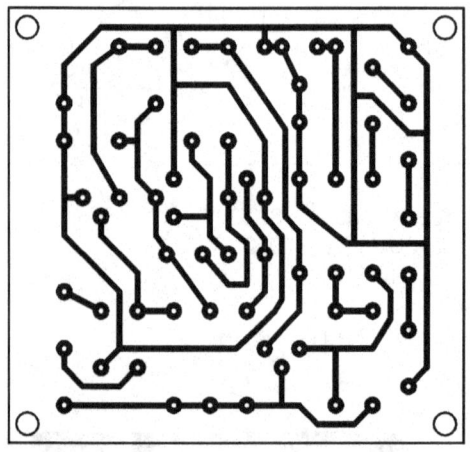

Figure 78.0 Printed Circuit Layout

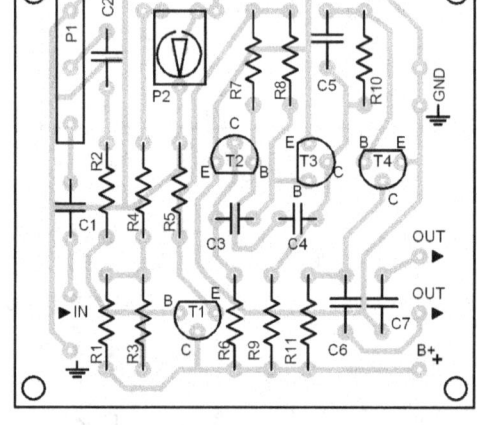

Figure 78.1 Parts Placement Layout

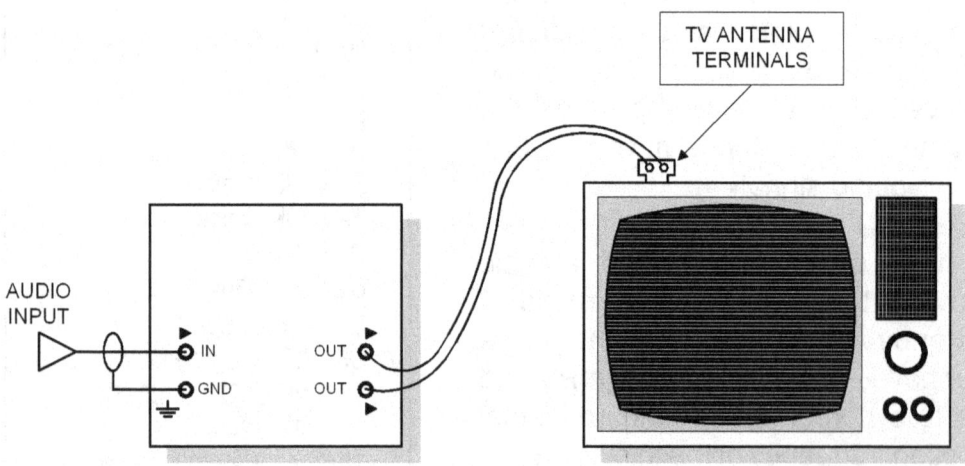

Figure 78.2 External Wirings

79 3-PHASE TESTER

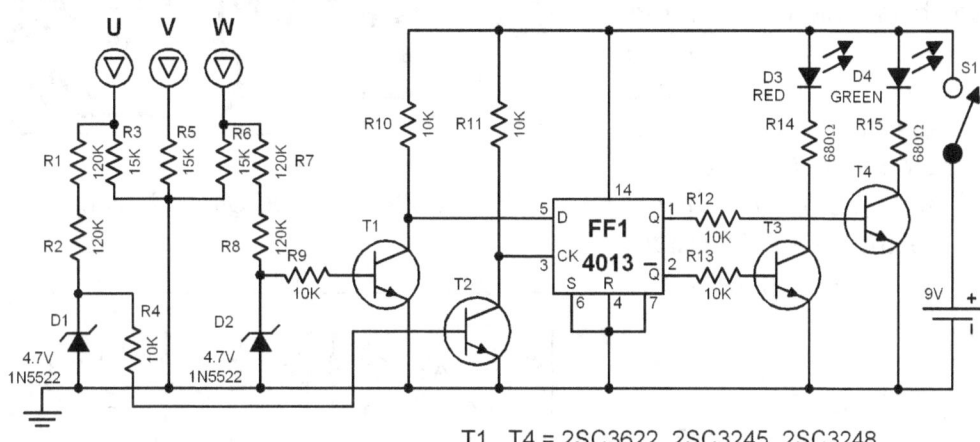

T1...T4 = 2SC3622, 2SC3245, 2SC3248

Diagram 79.0 3-Phase Tester

In connecting 3-phase motors to a 3-phase line, it is very important that the motor terminals are correctly connected to their corresponding line terminals. When it happens that the terminals are interchanged, the motor will rotate in the wrong direction. This could have dangerous and destructive consequences specially when the motor is installed in a machine. To avoid connecting the terminals in a wrong way, determine first the polarity of the line terminals. The circuit featured here can determine which should be connected to what.

3-phase motor terminals are designated as U,V and W. Accordingly, the power line terminals are marked with R,S and T.

The correct connection is as follows:

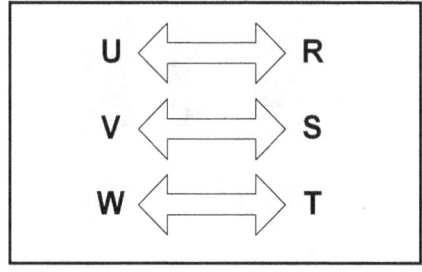

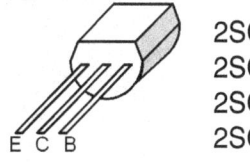

2SC3622
2SC3245
2SC3245A
2SC3248

The problem lies on the power line terminals. They are normally unmarked (cursing those lazy electricians wouldn't help either). This small tester finds out which is which. The three clips of the tester are marked U,V and W. These three clips are then connected to the power line. If the connection is correct, the green LED will light. Otherwise, the red LED will light warning that the connection is wrong.

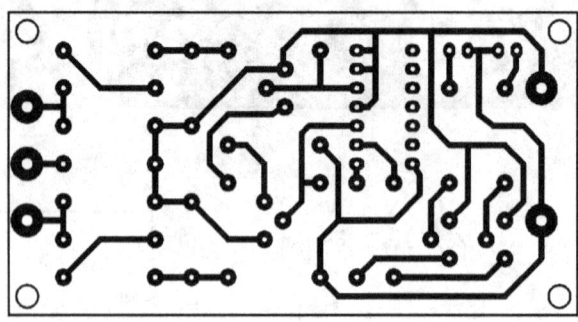

Figure 79.0 Printed Circuit Layout

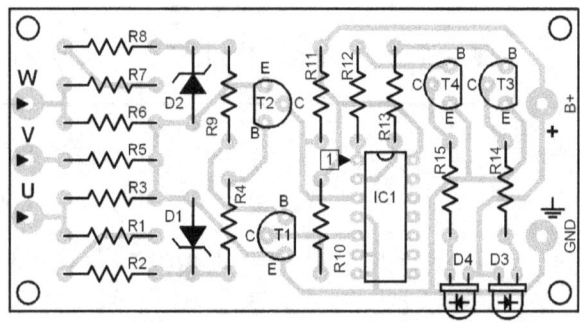

Figure 79.1 Parts Placement Layout

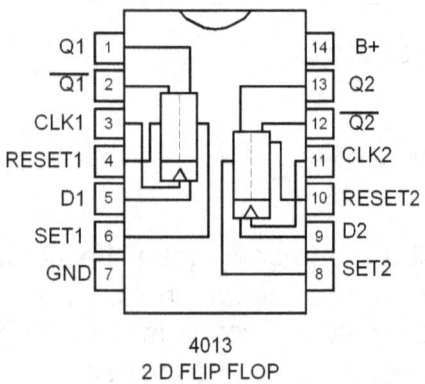

4013
2 D FLIP FLOP

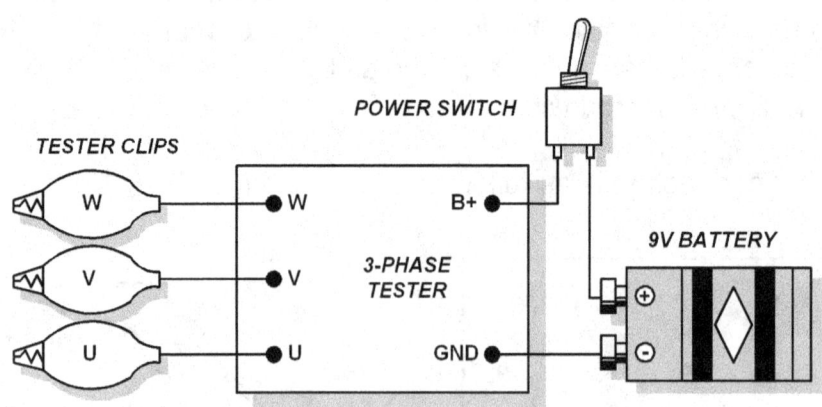

Figure 79.2 External Wirings

80 ZENER DIODE TESTER

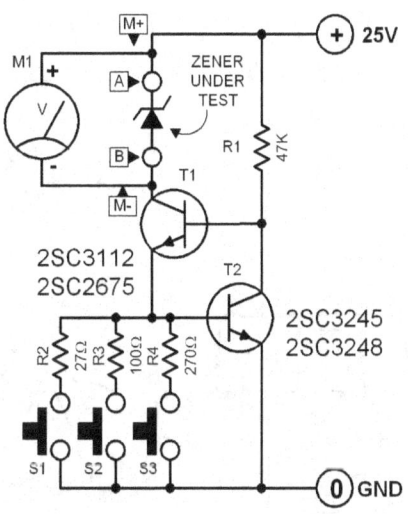

Diagram 80.0 Zener Diode Tester

This tester circuit is designed to determine the zener voltage of a zener diode that has no markings. The zener diode to be tested is inserted into test sockets (or clips), and a switch or a combination of switches S1...S3 is pressed. The zener voltage is displayed by the meter.

Switches S1...S3 causes different current values to flow through the diode. Table 80.0 shows the current value for every switch combination. Most zener voltages can be accurately tested with a current value of 5 mA or 10 mA (key S2 or keys S1+S2).

KEYS	V	Iz
S1	25V	2.22 mA
S2	25V	6 mA
S3	25V	22.2 mA
S1 + S2	25V	8.2 mA
S1 + S3	25V	24.2 mA
S2 + S3	25V	28.2 mA
S1 + S2 + S3	25V	30 mA

Table 80.0

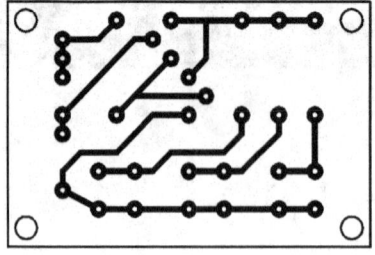

Figure 80.0
Printed Circuit Layout

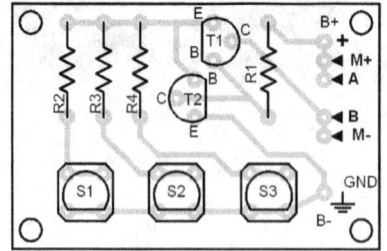

Figure 80.1
Parts Placement Layout

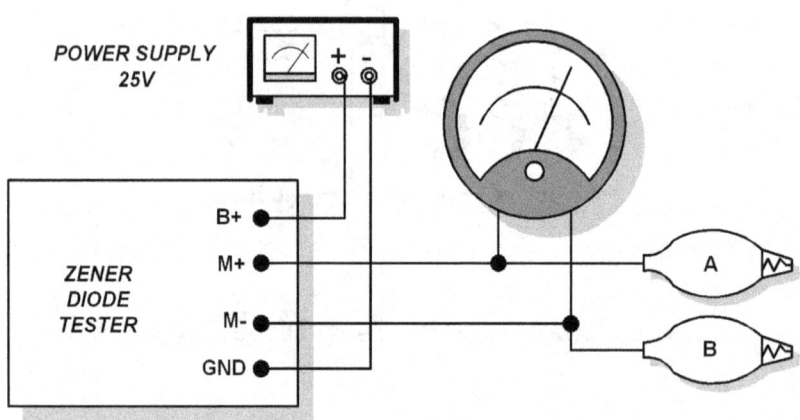

Figure 80.2 External Wirings

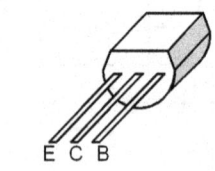

E C B

2SC3622   2SC2675
2SC3245   2SC3112
2SC3245A   2SC3248

Parts List:
R1 = 47K
R2 = 27Ω
R3 = 100Ω
R4 = 270Ω
T1 = 2SC2675
2SC3112
T2 = 2SC3622
2SC3245
2SC3248

DIGITAL & COMPUTERS

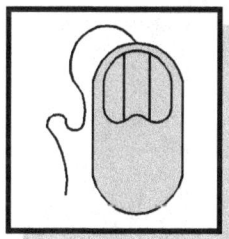

81   DIGITAL CLOCK

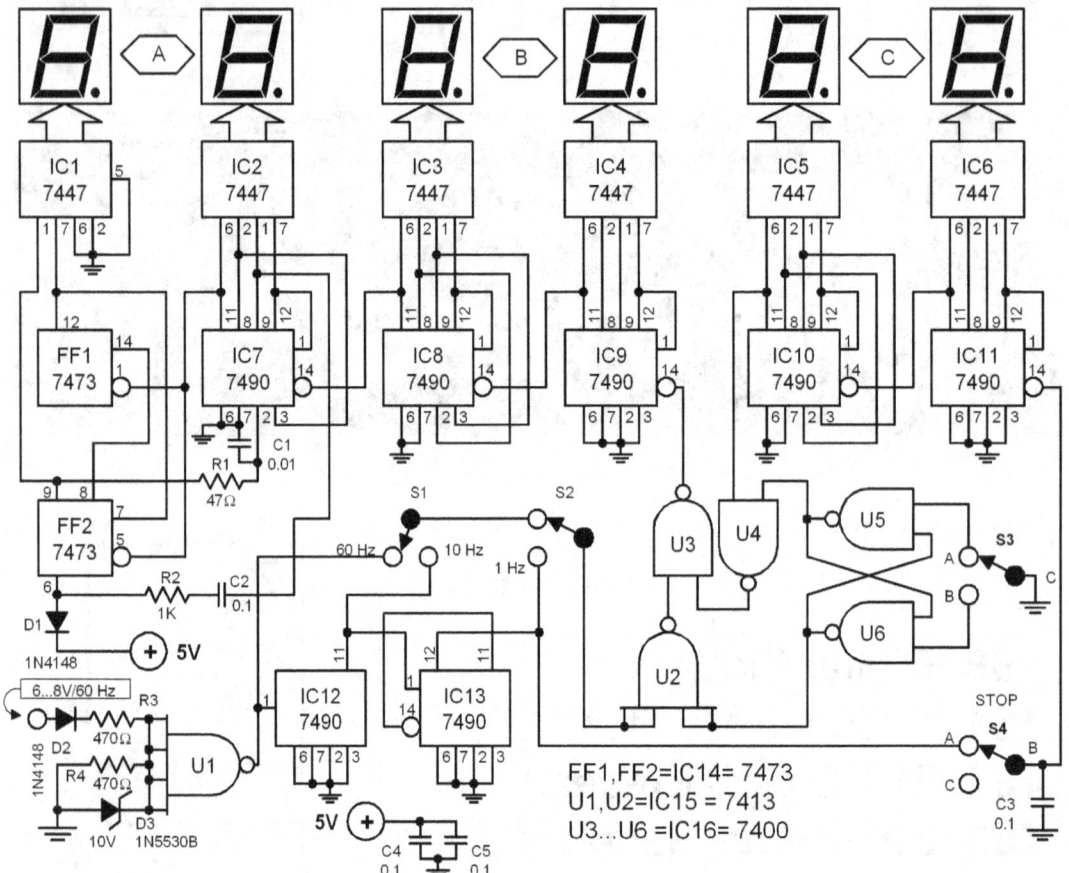

Diagram 81.0   Digital Clock

This clock circuit is constructed out of standard ICs. To set the time follow the instructions below:

a) Opening S3 stops the clock
b) Switch S1 selects between fast and slow setting of time
c) Pressing S2 drives the clock with a much higher speed.
     When the desired time setting is reached, release S2.
d) Closing S3 starts the clock.

Note: This clock circuit works in a 24-hour mode.

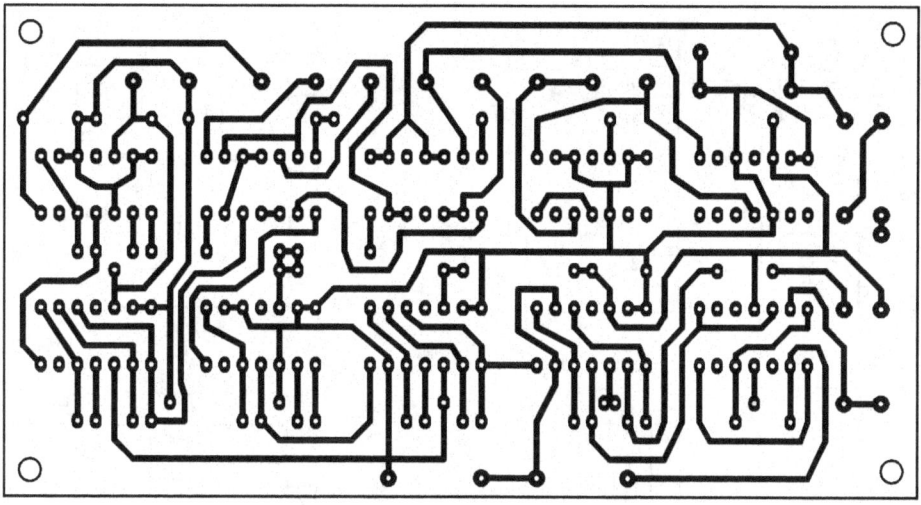

Figure 81.0 Printed Circuit Layout

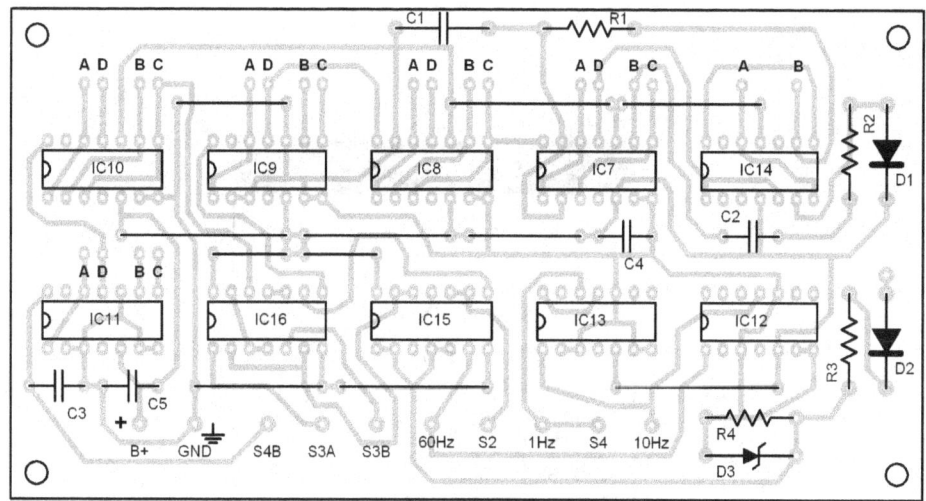

Figure 81.1 Parts Placement Layout

7400
TTL QUAD NAND

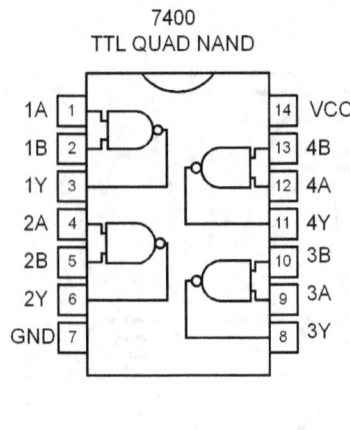

7413
TTL 4-INPUT SCHMITT TRIGGER

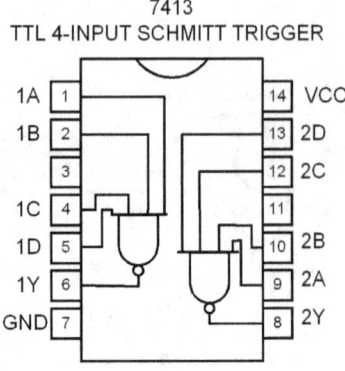

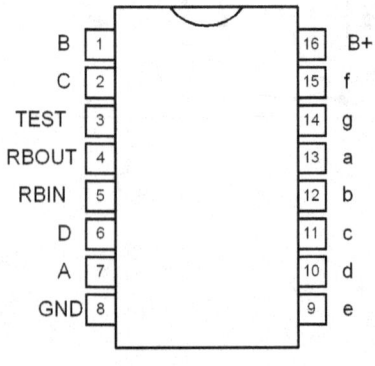

7447
TTL BCD-7SEGMENT
DECODER/DRIVER

7413
TTL 4-INPUT SCHMITT TRIGGER

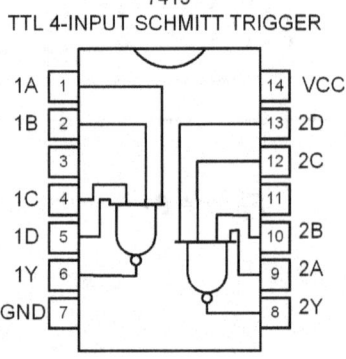

Parts List:

R1 = 47Ω
R2 = 1K
R3,R4 = 470Ω
C1 = 0.01/50V
C2,C3,C4,C5 = 0.1/50V
D1,D2= 1N4148
D3 = 1N5530B(10V)
IC1...IC6 =   7447
IC7...IC13 = 7490
IC14 = 7473
IC15 = 7413
IC16 = 7400

82 INTERFACE MONITOR

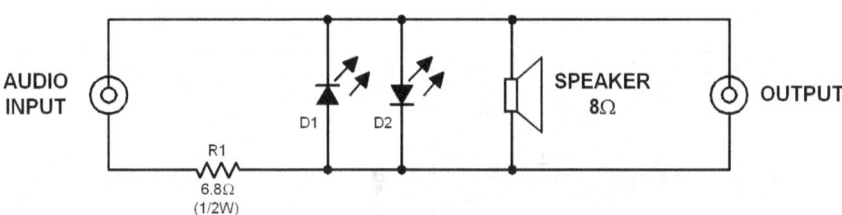

Diagram 82.0 Interface Monitor

This small circuit provides simultaneous optical and acoustic signals for monitoring the digital data being relayed from the cassette recorder to the microcomputer and vice versa. Most cassette interfaces need a signal level of around 2 volts to be able to read the data correctly.

Once the LEDs light continously, it means that the output signal is too strong. The volume control must be turned back a little until the LEDs flicker only slightly. The two LEDs are wired antiparallel. The loudspeaker functions as an additional monitor so that you will know when the recorder delivers data (you can hear two different tones) or in a pause between two programs (you can hear the tape noise only).

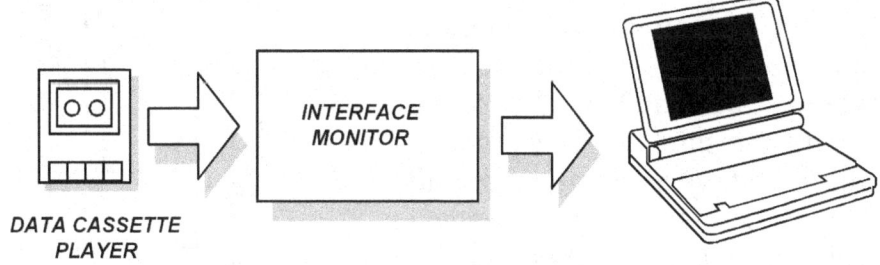

DATA CASSETTE
PLAYER

Figure 82.0 Application of Interface Monitor

83   EPROM IC ERASER

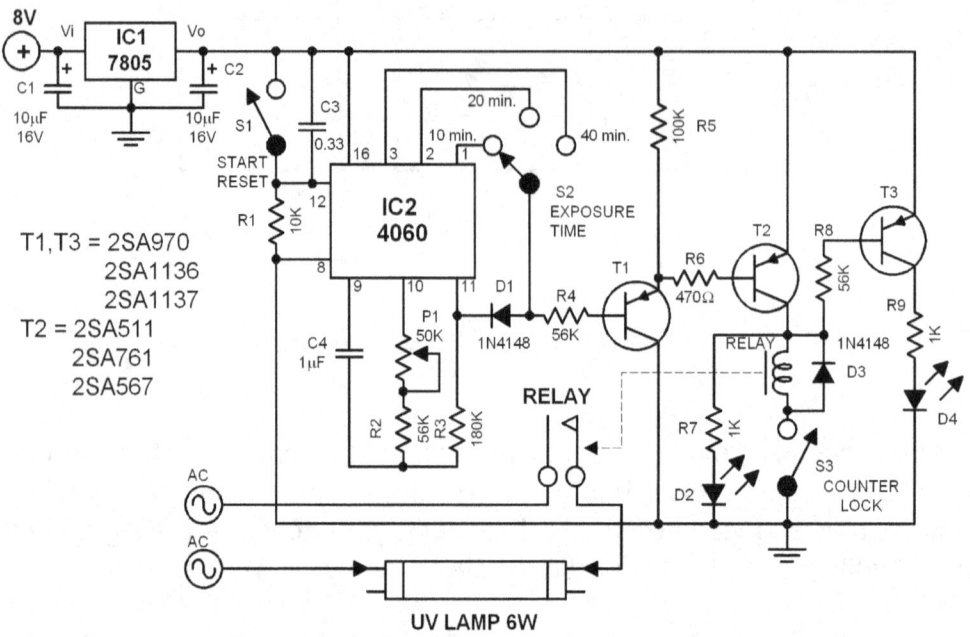

Diagram 83.0   EPROM IC Eraser

It is very easy to erase an EPROM, one needs to switch on an ultraviolet lamp and place it long enough (10-40 minutes depending on the manufacturer's specifications) around 2cm above the EPROM and it's finished. Well, the problem is that it is not very comfortable to continously watch the clock while waiting until the erasure process is completed. UV rays are also dangerous to the eyes. Why not use electronics to automate the whole job?. You are working with electronics anyway, aren't you?

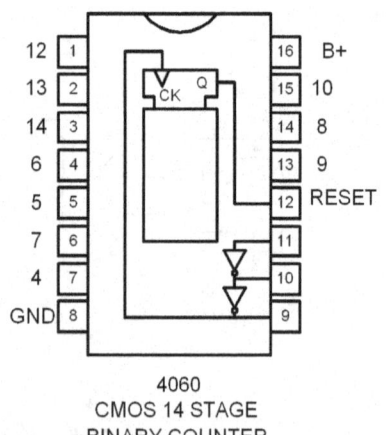

4060
CMOS 14 STAGE
BINARY COUNTER

The circuit featured here controls the "exposure time" automatically and signals when the erasure is completed. When S1 is closed, the relay is activated and the UV lamp lights. At the same time, D2 (red) lights showing that erasure is taking place. When the erasing (exposure) time has elapsed (time is selectable through S2), the relay deactivates and the UV lamp, together with D2, goes off. At this time, D3 lights signalling that the job is finished.

In constructing the circuit, it is important that the UV lamp is contructed inside a nontransparent box. Switch S3 is a counterlock and must be installed inside the box in such a way that it is switched on only when the box is closed. Lifting the box cover opens this switch automatically and the relay releases switching off the UV lamp. This technique ensures that the UV lamp lights only when the box is already closed preventing the UV rays from radiating out. The best type for S3 is a microswitch - see the illustration. The switch S1 resets the counter.

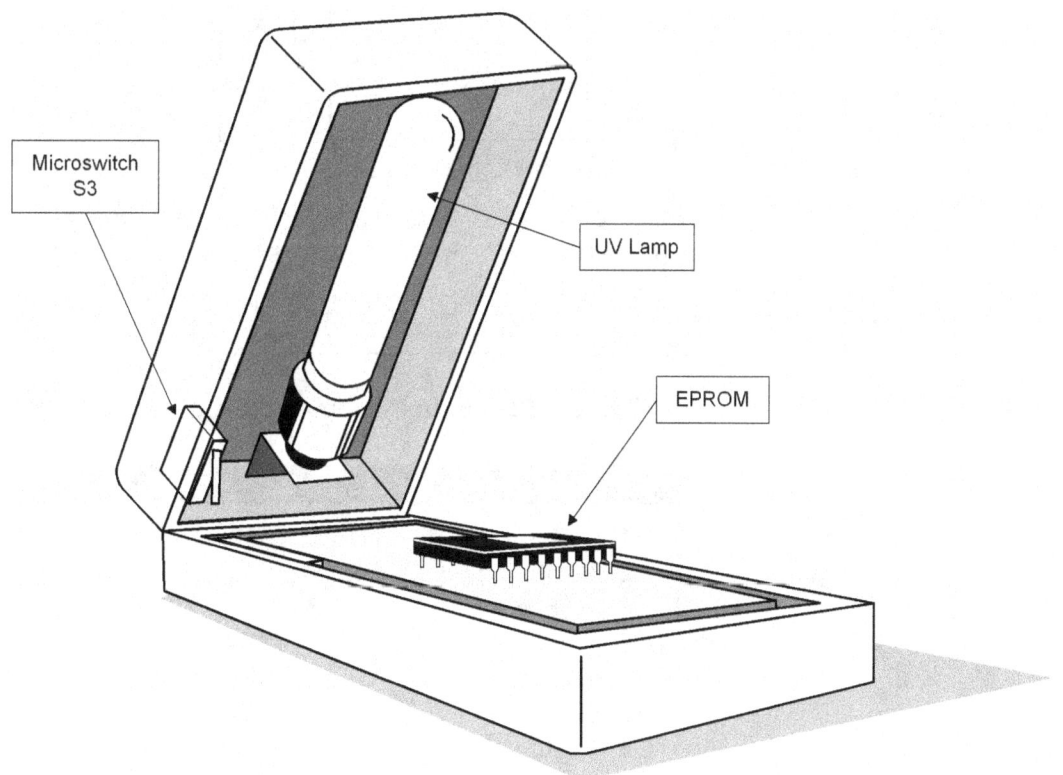

Figure 83.0 Construction Example

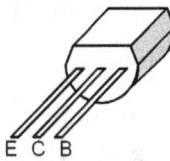

2SA970
2SA1136
2SA1137

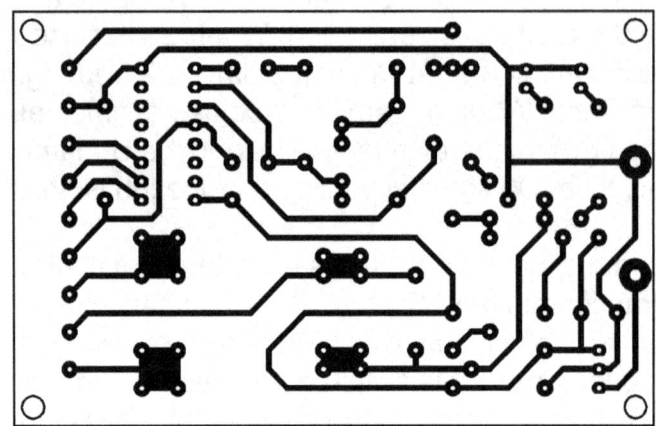

Figure 83.1 Printed Circuit Layout

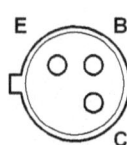

2SA511
2SA567
2SA761

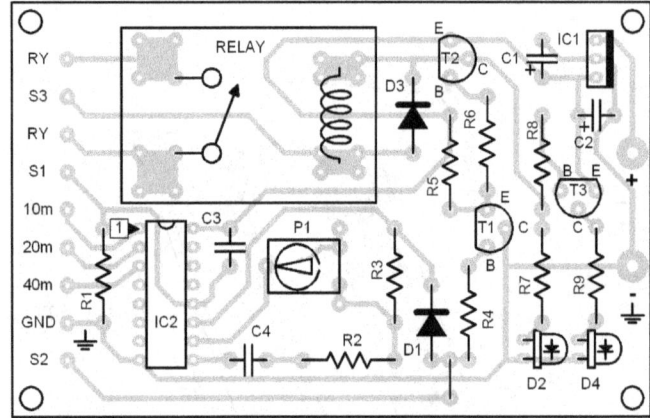

Figure 83.2 Parts Placement Layout

7805
5-Volts
voltage regulator

Parts List:	
R1 = 10K	D1,D3 = 1N4148
R2,R4,R8 = 56K	D2,D4 = LED
R3 = 180K	T1,T3 = 2SA970(2SA1136)
R5 = 100K	(2SA1137)
R6 = 470Ω	T2 = 2SA511(2SA567)
R7,R9 = 1K	(2SA761)
C1 = 0.1/50V	IC1 = 4060
C2 = 10µF/25V	
C3 = 0.33/50V	6 watts Ultraviolet lamp
C4 = 1µF/25V	S3 = microswitch (SPSP)

84 BAUD RATE GENERATOR

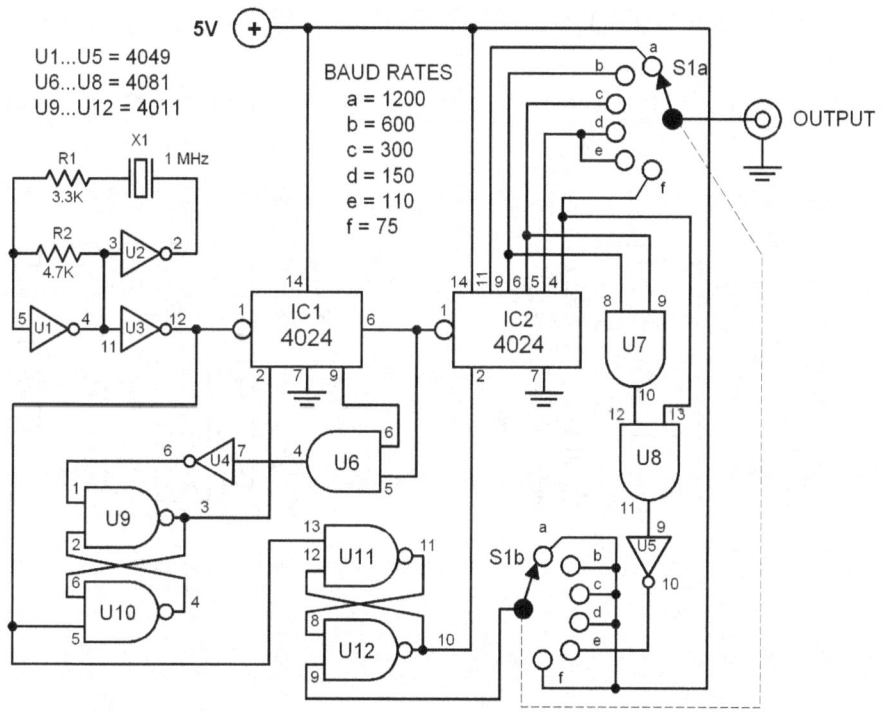

Diagram 84.0 Baud Rate Generator

This switchable oscillator/counter functions as a baud rate generator and can generate transmission speeds of 1200, 600, 300, 150, and 75 baud. This circuit is specially designed to provide an external clock signal for transmitting asynchronous binary data from asynchronous transmitter/receivers which normally function with clock speeds 16 times faster than the transmission rates.

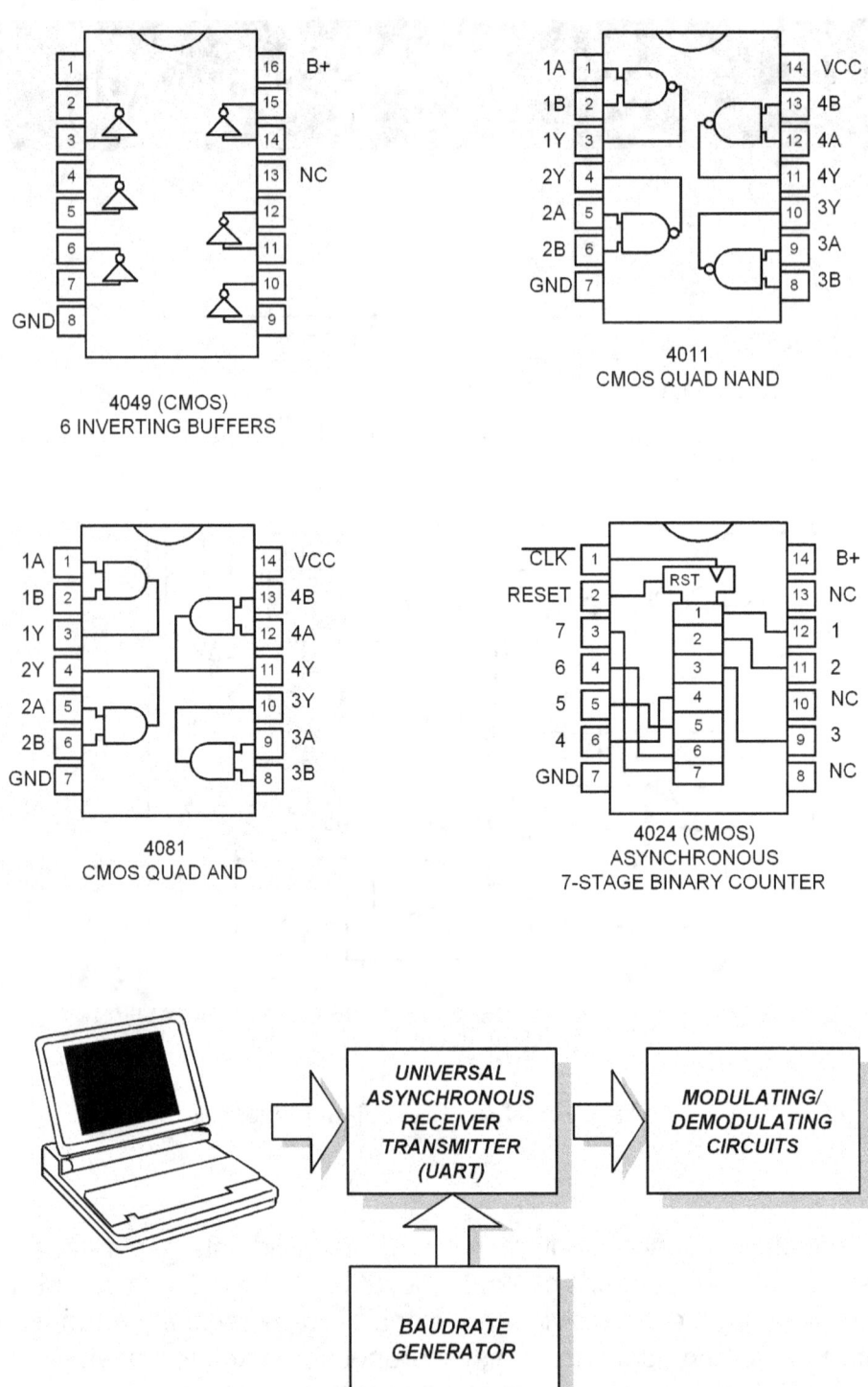

4049 (CMOS)
6 INVERTING BUFFERS

4011
CMOS QUAD NAND

4081
CMOS QUAD AND

4024 (CMOS)
ASYNCHRONOUS
7-STAGE BINARY COUNTER

Figure 84.0   Typical Application of the Baudrate Generator

85 DATA MONITOR

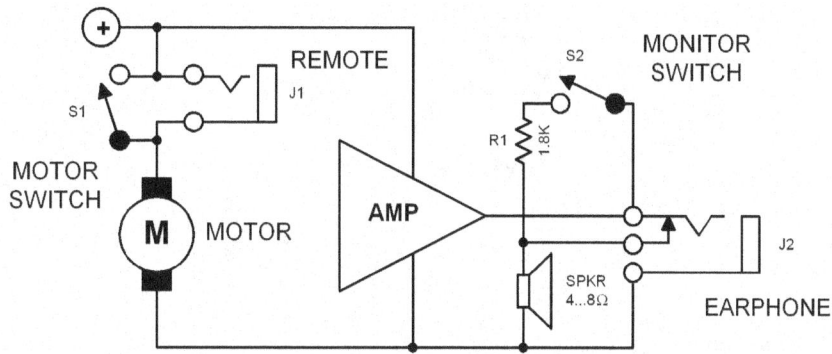

Diagram 85.0 Data Monitor

This simple circuit connections can actually help you in your work in transferring programs from cassettes to the floppy. Normally the cassette interface is inserted into the speaker output of the recorder which disables the speaker once the interface plug is inserted. The problem arises when you want to find a certain program. First, you have to unplug the interface, play the cassette and then plug the interface back when the program is found.

The circuit featured here offers a fast and convenient way. Switching S2 connects a speaker to the output in parallel. This way, you can monitor the program (in audio of course) without removing the interface plug. The circuit has another switch to control the motor of the player through its remote line. Since remote terminals differ with every brand of player, only a concept diagram is shown here.

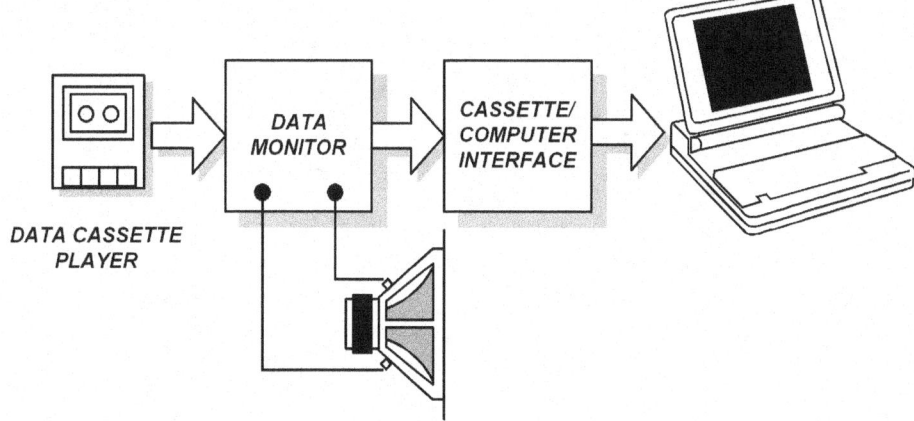

86 80-40 TRACK CONVERTER

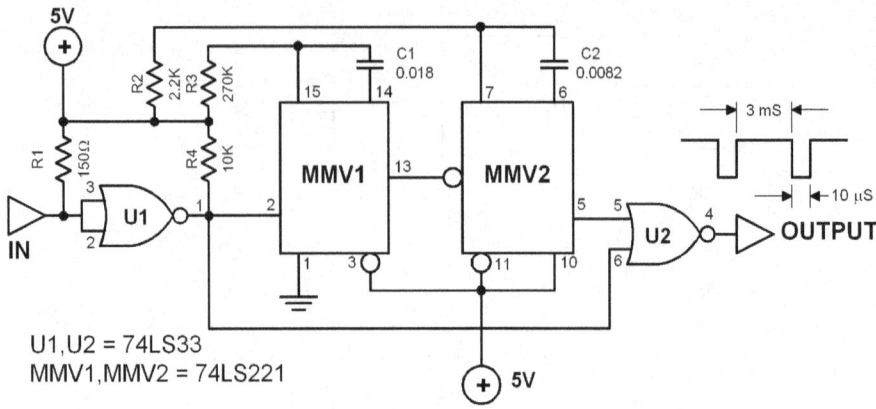

Diagram 86.0 80-40 Tracks Converter

This circuit causes a "new" 80-track diskdrive to act like your faithful "old" 40-tracker. This enables your computer to read data from a diskette with a 40-track format. The 40-track data is twice as wide as the 80-track data that is why it cannot be read anymore by an 80-track diskdrive.

By using this circuit however, the read/write head of the 80-tracker is "pushed" two tracks further everytime a control pulse is delivered by the computer. With this technique, the 80-track diskdrive can read the 40-track data.

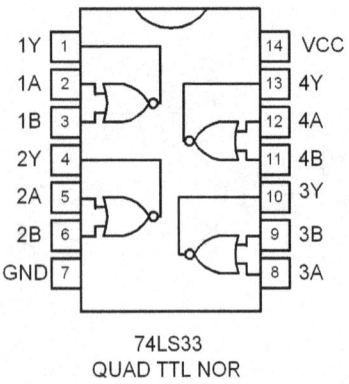

74LS33
QUAD TTL NOR

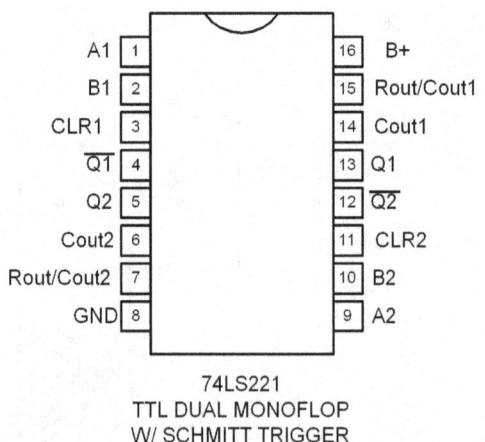

74LS221
TTL DUAL MONOFLOP
W/ SCHMITT TRIGGER

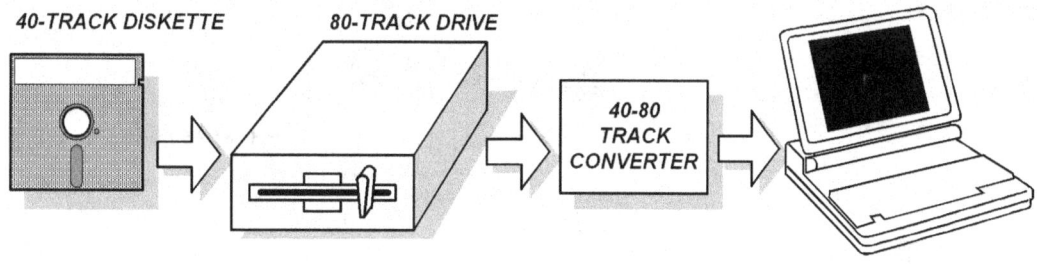

Figure 86.0   Sample Application of the Converter

87   A/D & D/A INTERFACE

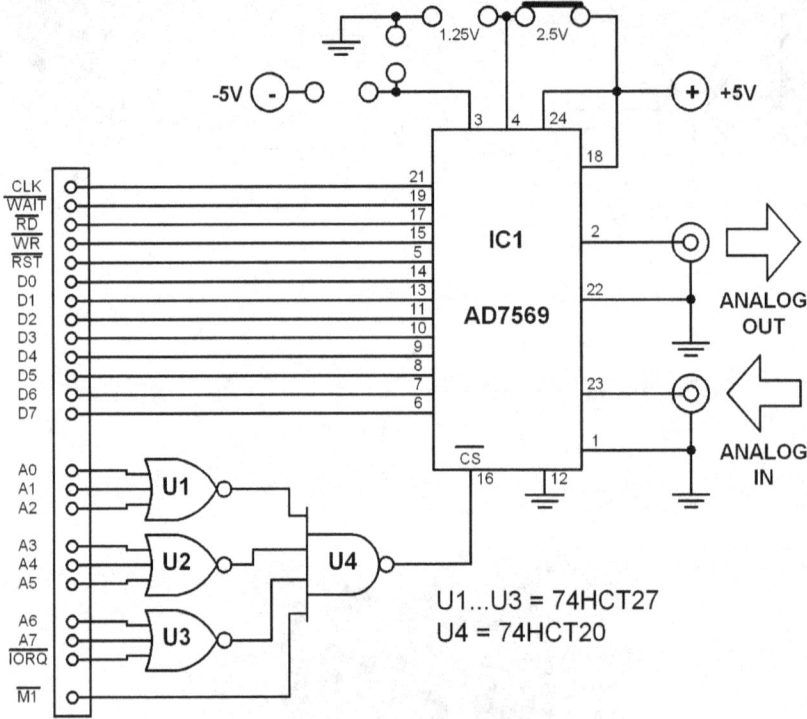

Diagram 87.0   A/D & D/A Interface

This circuit uses a single chip analog I/O system AD7569 from Analog Devices. This highly integrated chip contains an 8-bit A/D converter with 2 μs conversion time and a D/A converter with 1μs conversion time. Reference voltages and a bus interface for direct-to-computer connection are also integrated. The address coding is, however, not contained in the chip. Gates U1...U4 decode the lower 8 address lines for the read/write operations.

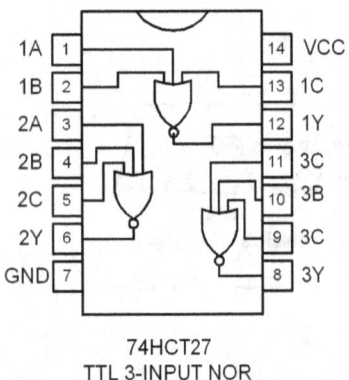

74HCT27
TTL 3-INPUT NOR

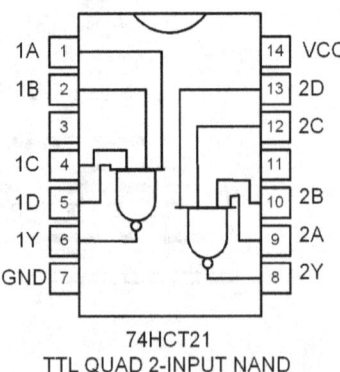

74HCT21
TTL QUAD 2-INPUT NAND

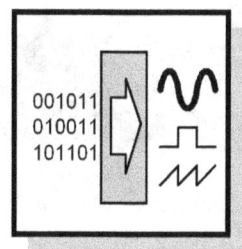

OSCILLATORS & COUNTERS

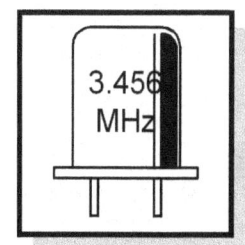

3.456 MHz

88 ONE CHIP VCO

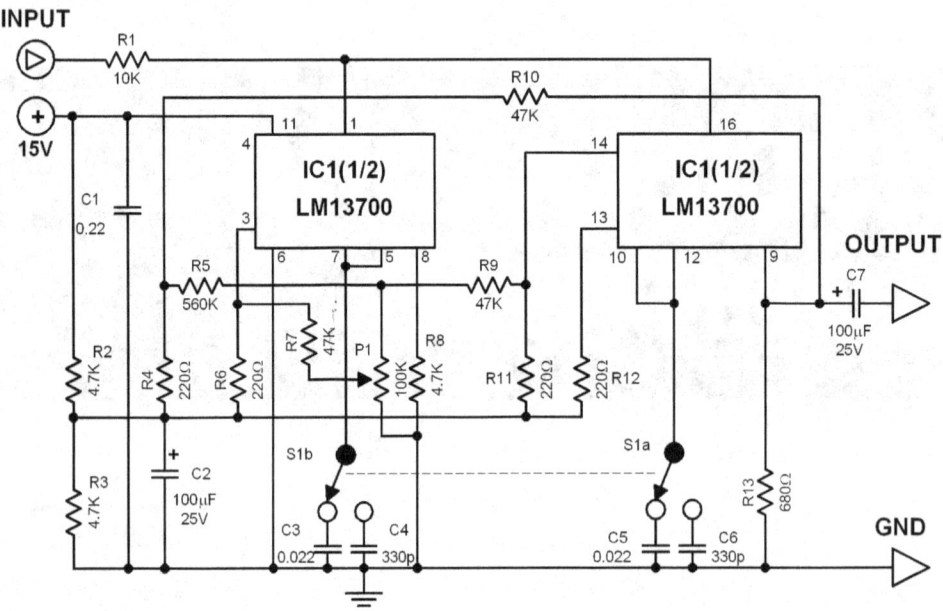

Diagram 88.0   One Chip VCO

This is a sinewave oscillator generating a frequency that is controllable by a DC voltage. The oscillator has two frequency ranges: 6.7Hz to 400 Hz and 400 Hz to 23.8 kHz selectable through S1. The distortion factor is less than 1% when the amplitude is under 10 volts p-p. This distortion sinks to around 0.15 % when the output voltage is set to 1 volt p-p. The IC loses its temperature insensitivity and stability when the output is lower than 1 volt p-p. The output current comes from pins 5 and 12. This current is controlled through the input pins 1 and 16 and can reach a maximum of 0.75 mA.

Potentiometer P1 varies the output frequency. The control voltage can vary between 1.3 volts and 15 volts in covering the entire frequency range. The supply voltage can also affect the output frequency. To produce a constant frequency, use a well regulated supply to power the circuit. The current consumption is around 6.5 mA.

89 2-FUNCTION COUNTER

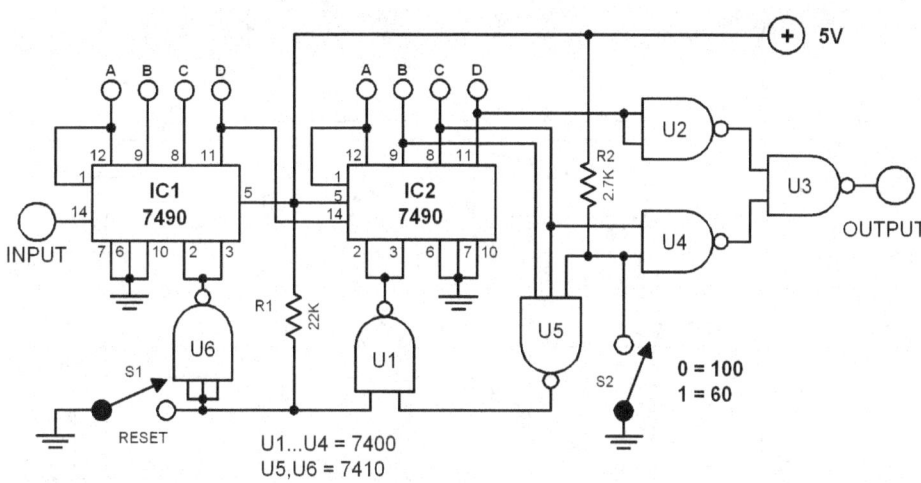

Diagram 89.0 2-Function Counter

This circuit is designed as a compact module enabling it to be cascaded to obtain the desired counting range. This counter module has a double function. First, it can divide by 100 which is usually needed in decimal systems. Second, it can divide by 60 which is important for clock circuits counting in seconds and minutes. The division factor is selectable through S2.

If you cascade two identical modules, you will get a 4-decade counter which is convertible for time measurements. Cascading more modules is of course possible.

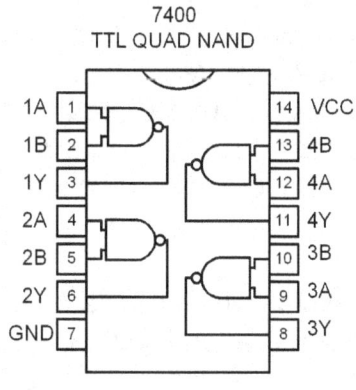

7400
TTL QUAD NAND

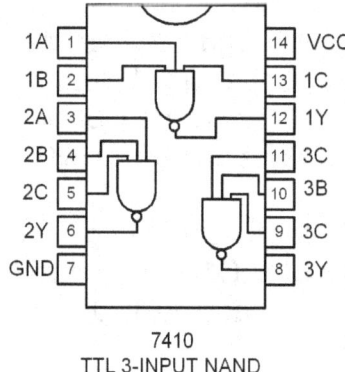

7410
TTL 3-INPUT NAND

90 PLL OSCILLATOR

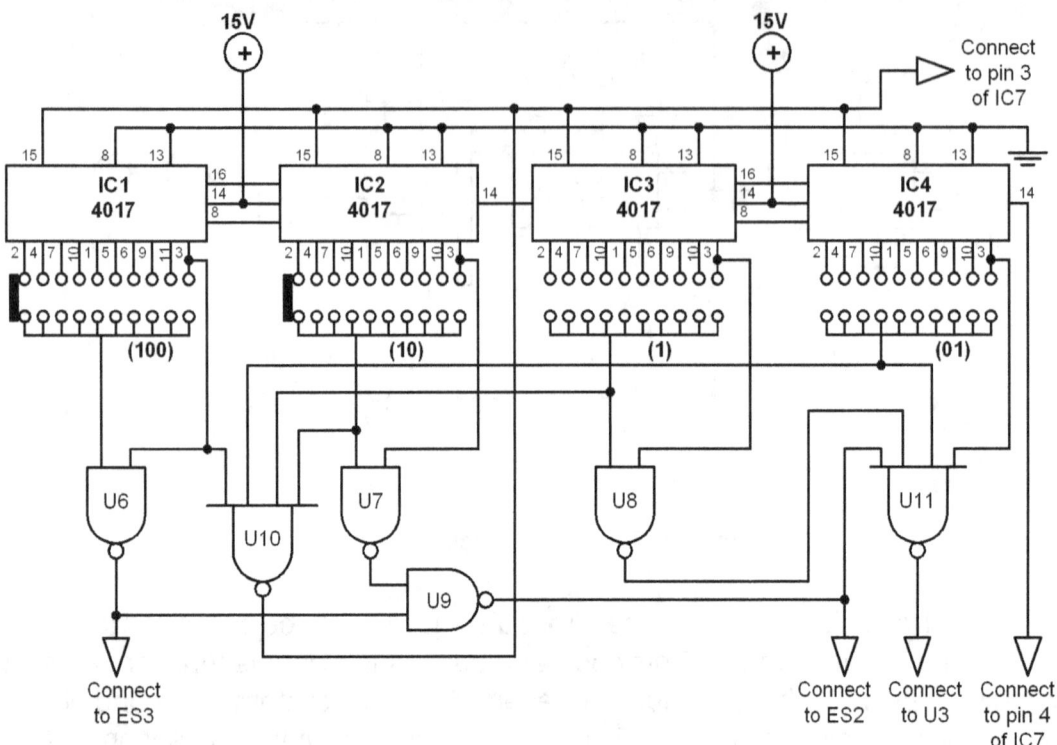

Diagram 90.0 PLL Oscillator

This digitally controlled crystal oscillator is a frequency synthesizer designed to produce low frequency squarewave signals. The stability of the generated signal depends on the stability of the crystal being used. The generated signal is processed by four main circuits which function together as phase locked loop (PLL). Depending on the needs of the user, the generated signal can also be channeled through a divider which divides it by 1000. A buffer circuit at the final stage allows the signal to be connected to either TTL or CMOS compatible circuits. The signal frequency is selectable through four switches (S3..S6) between 0.4 Hz and 999.9 kHz.

The main oscillator is being controlled by the crystal which oscillates at 3.2768 MHz. The ICs IC12... IC14 function as the 1:1000 divider. LED D1 lights when the PLL is already "locked" - meaning the desired frequency is already being generated.

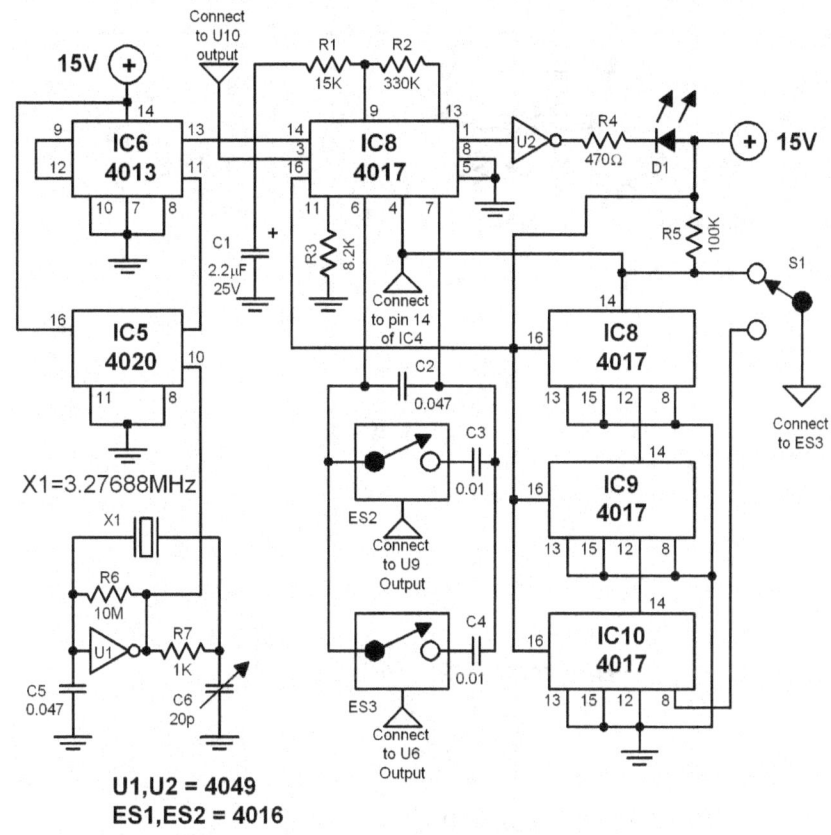

Diagram 90.1 PLL Oscillator

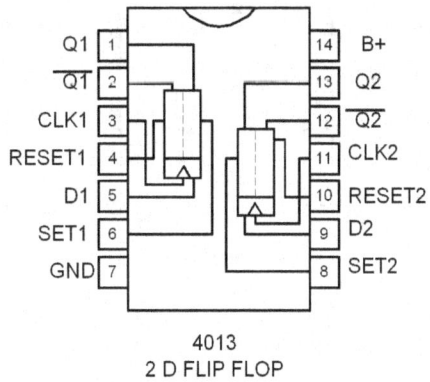

4013
2 D FLIP FLOP

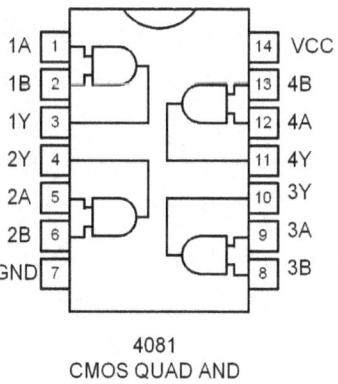

4081
CMOS QUAD AND

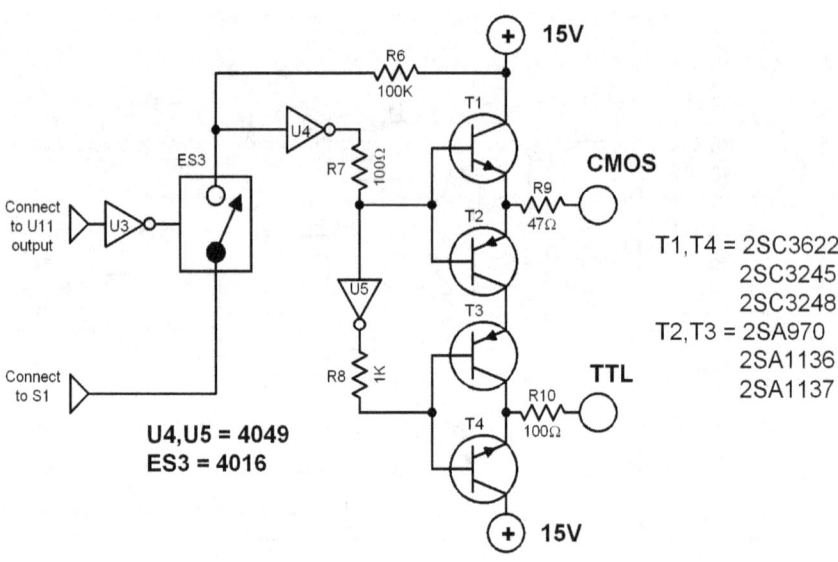

Diagram 90.2   PLL Oscillator

2SC3622	2SA970
2SC3245	2SA1136
2SC3248	2SA1137

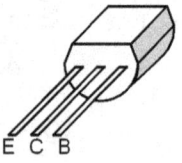

E C B

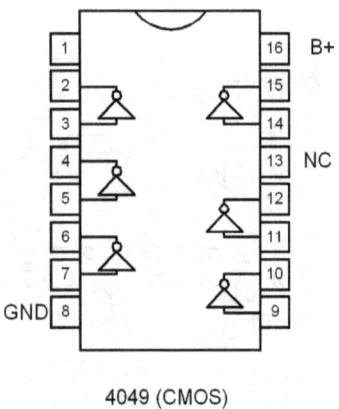

4049 (CMOS)
6 INVERTING BUFFERS

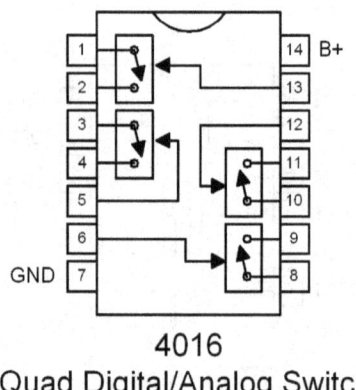

4016
Quad Digital/Analog Switch

91 SINEWAVE OSCILLATOR

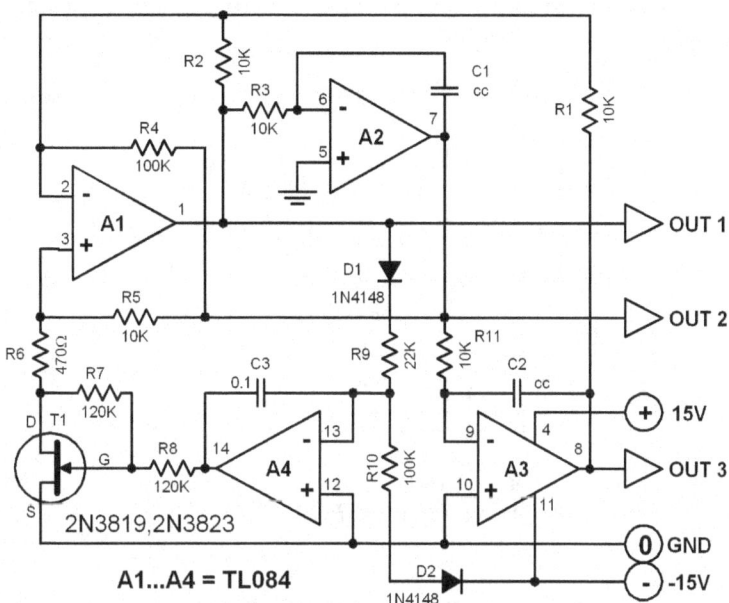

Diagram 91.0 Sinewave Oscillator

A sinewave oscillator can be constructed by coupling the output of a selective filter circuit back to its input as long as certain rules are followed. This technique is applied in the featured circuit. This circuit is a variable-state-filter. Its amplitude is being stabilized by a voltage controlled FET transistor. The control voltage for the FET comes from the ouput of the amplifier U1. This control voltage flows first through the diode/ resistor network and the integrator U4 before reaching the FET. To find the values of C1 and C2 use the following formula:

$$C1 = C2 = \frac{16}{f}$$

where C1 and C2 = µF and f = kHz.

Each of the outputs 1,2 and 3 delivers a sine wave signal. Output 3 delivers the least distorted sinewave while output 1 delivers the most distorted sinewave.

92   NE5517 VCO

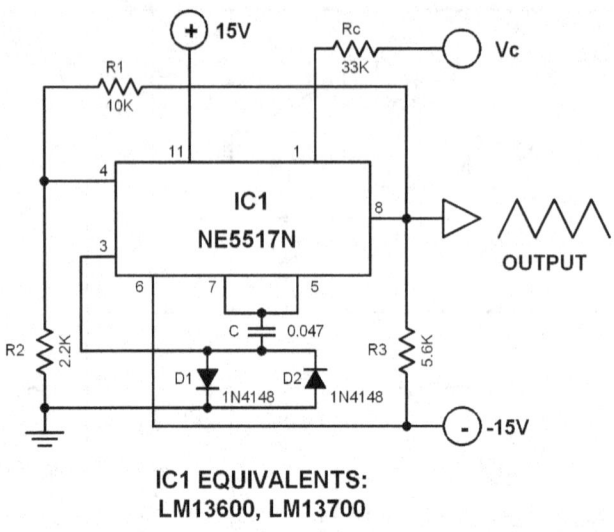

IC1 EQUIVALENTS:
LM13600, LM13700

Diagram 92.0   NE5517 VCO

The IC NE5517N is applied as a voltage controlled oscillator (VCO) in this circuit to generate a trianglewave signal. The integrated amplifier is countercoupled through the divider network R1 and R2. The capacitor C is almost linearly charged and discharged. The charging or discharging current flows through one of the diodes. The switchover level is around 0.59 volts.

The signal frequency **f** can be found with:

$$f(Hz) = \frac{Vc + 15}{2.4CRc}$$

The output amplitude **Va** can be found with:

$$Va(pp) = 1.2 \left(\frac{R1 + R2}{R2} \right)$$

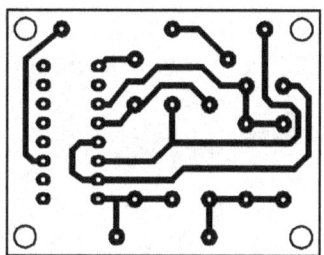

Figure 92.0 Printed Circuit Layout for NE5517 VCO

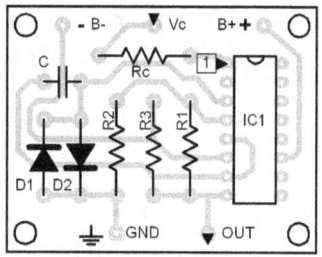

Figure 92.1 Parts Placement Layout for NE5517 VCO

93   100 MHz OSCILLATOR

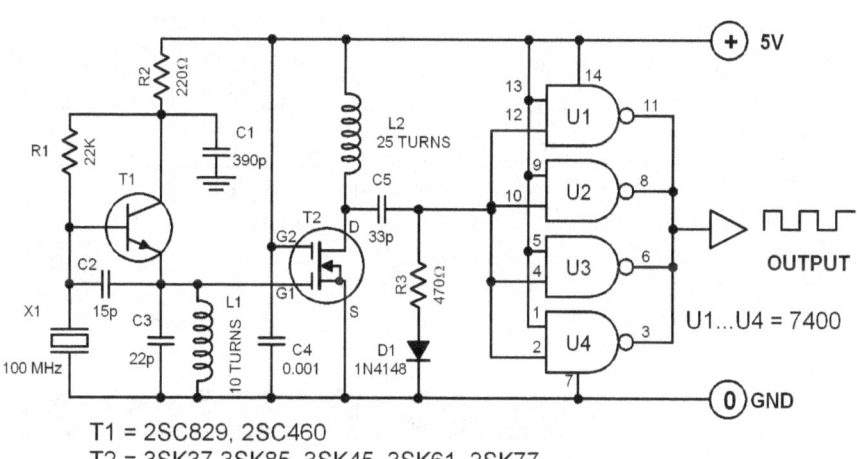

T1 = 2SC829, 2SC460
T2 = 3SK37,3SK85, 3SK45, 3SK61, 3SK77

Diagram 93.0   100 MHz Oscillator

This circuit produces a crystal-stabilized 100 MHz clock signal with a TTL compatible output. The crystal oscillates at its 5th overtone, and can be fine tuned by the L1/C2 combination. The crystal has a basic frequency of 20 MHz, and is excited to a frequency of 100 MHz through parallel resonance.

The dual gate MOSFET works as a buffer to prevent the oscillator from being loaded. L1 is 10 turns 0.5mm magnet wire and L2 is 25 turns 0.3mm magnet wire. Both coils have a coil diameter of 3 mm. All connecting wires must be as short as possible.

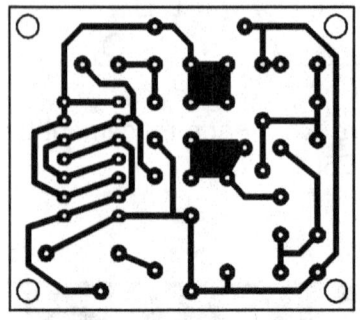

Figure 93.0 Printed Circuit Layout

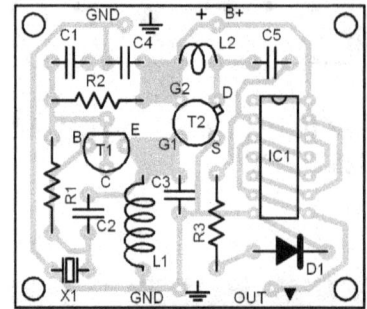

Figure 93.1 Parts Placement Layout

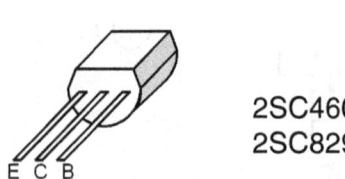

2SC460
2SC829

7400
TTL QUAD NAND

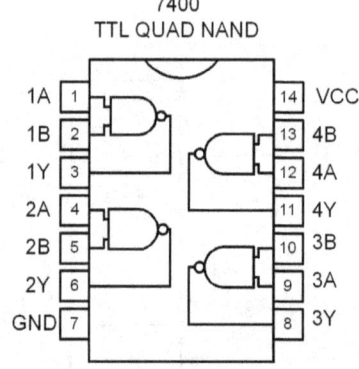

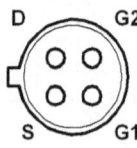

2SK37
2SK45
2SK61

Parts List:

R1 = 22K

R2 = 220Ω

R3 = 470Ω

C1 = 390p/50V

C2 = 15p/50V

C3 = 22p/50V

C4 = 0.001/500V

C5 = 33p/50V

D1 = 1N4148

T1 = 2SC460 (2SC829)

T2 = 3SK37 (3SK45)
 (3SK61)

IC1 = 7400

X1 = 100 MHz crystal

AUXILIARY

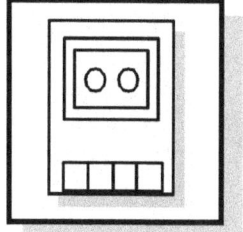

94   MORSE KEYER

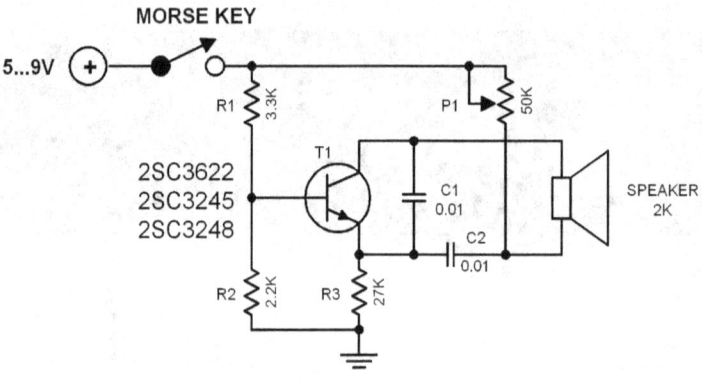

Diagram 94.0   Morse Keyer

This morse keyer is used in practicing morse code transmission and is very helpful in preparing for a license examination. The tone frequency can be varied through R4.

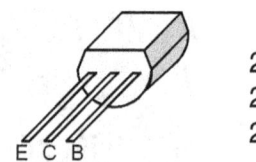

2SC3622
2SC3245
2SC3249

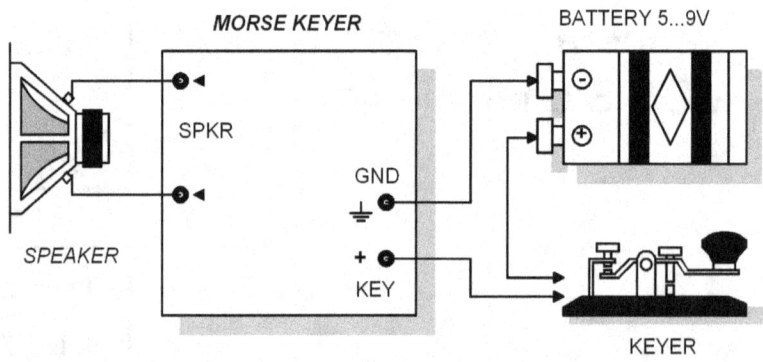

Figure 94.0   External Wirings

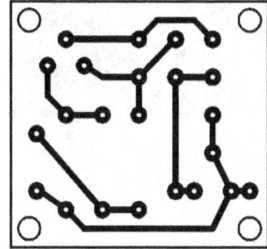

Figure 94.1 Printed Circuit Layout

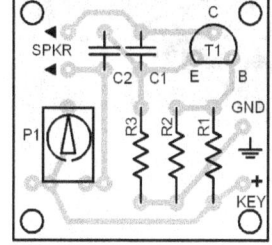

Figure 94.2 Parts Placement

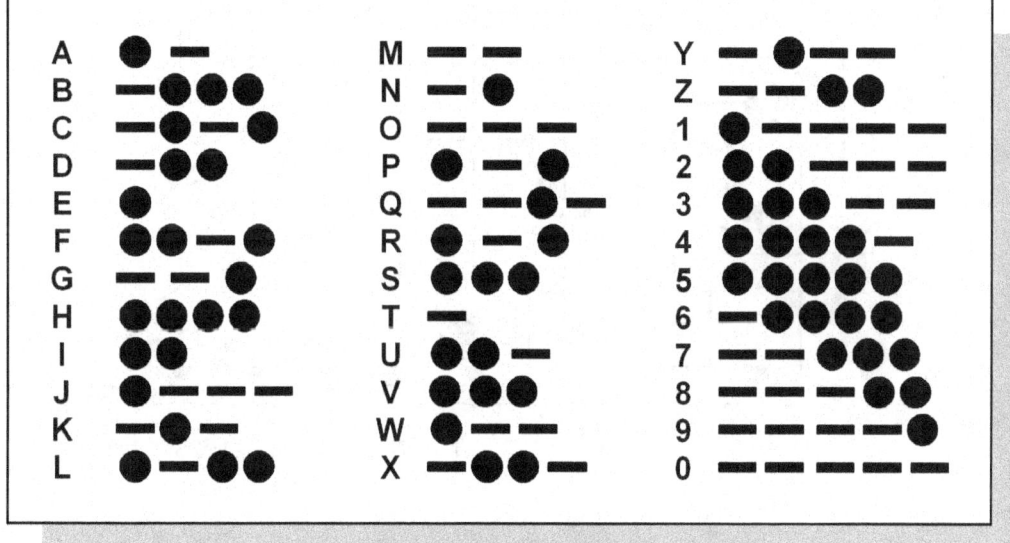

Table 94.0 Morse Code

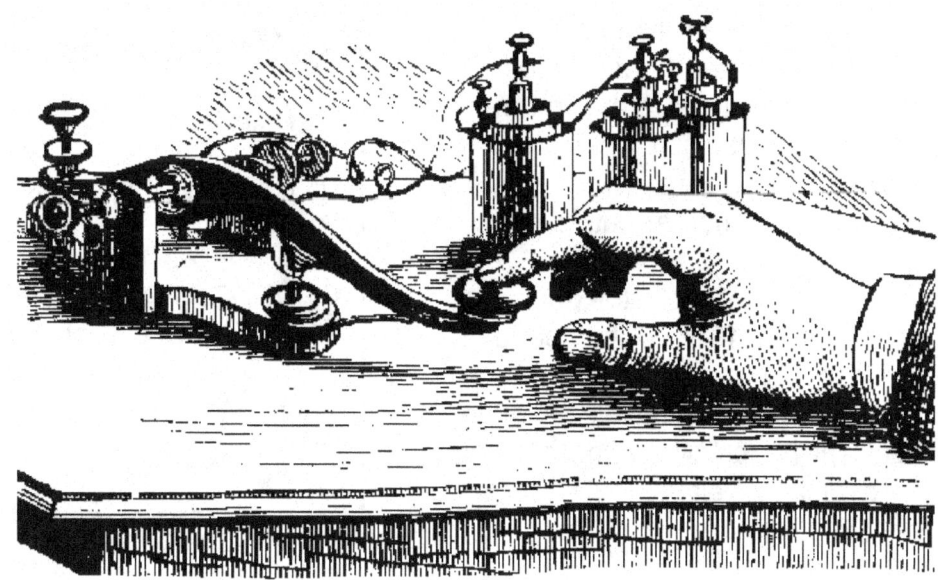

95   RUNNING LIGHT

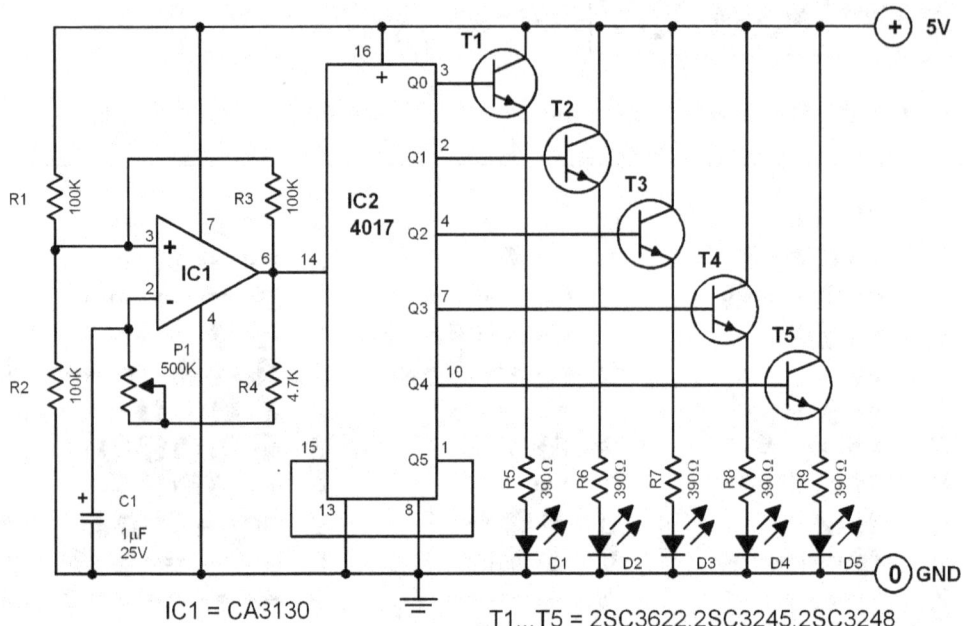

Diagram 95.0 Running Light

This circuit drives a series of LEDs to blink in a certain pattern that creates an illusion of a running light. A maximum of 10 LEDs can be used. If more than 5 LEDs are used, then the reset pin 15 must be reconnected to another output pin.

IC1 functions as an oscillator driving a decimal counter IC 4017. The speed of the running light can be varied through P1. The outputs can be used also to drive higher current lamps by adding current amplifier circuits or relays.

Parts List:

R1,R2,R3 = 100K
R4 = 4.7K
R5,R6,R7,R8,R9 = 470Ω
P1 = 500K
C1 = 1µF/25V
D1,D2,D3,D4,D5 = LED
T1,T2,T3,T4,T5 = 2SC3622
      (2SC3245)(2SC3248)
IC1 = CA3130
IC2 = 4017

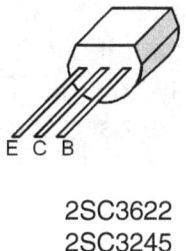

2SC3622
2SC3245
2SC3249

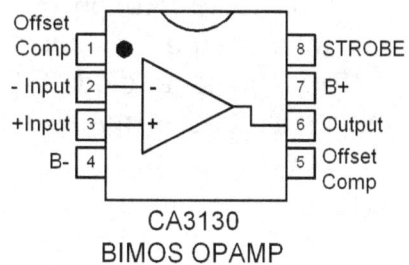

CA3130
BIMOS OPAMP

96 °C/VOLTAGE CONVERTER

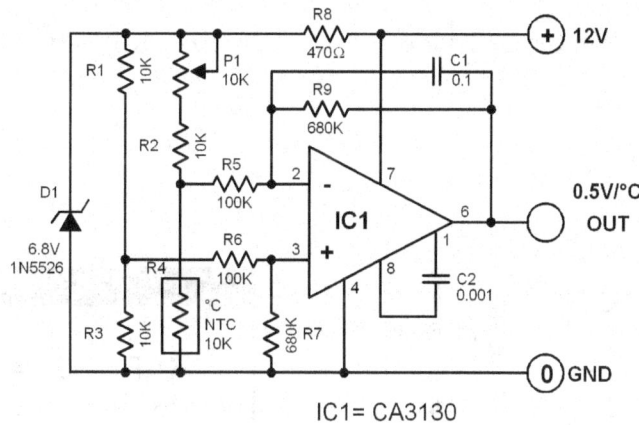

IC1= CA3130

Diagram 96.0 °C/Temperature Converter

This circuit can convert temperature level (in degree Centigrade) into a voltage value. The conversion factor is 0.5V per °C. An NTC type resistor is used for the temperature sensor. The resistance of NTC type resistors is well known to be dependent on the outside temperature. With this circuit, you can measure the room' s temperature accurately.

The converter is basically a bridge circuit that is connected to a regulated power supply. It is calibrated through P1 so that the output is zero at the lowest measurable temperature. The supply voltage is not critical. It can be between 4.7V and 12V. The zener diode D1 regulates the supply voltage. The current consumption is around 11 mA. A digital voltmeter can be used to display the measured value.

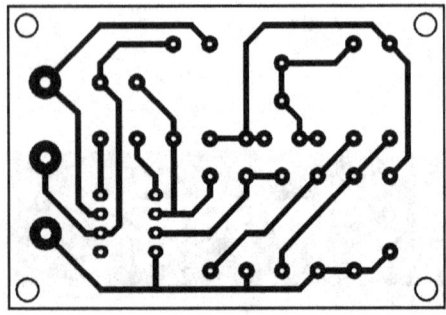

Figure 96.0 Printed Circuit Layout

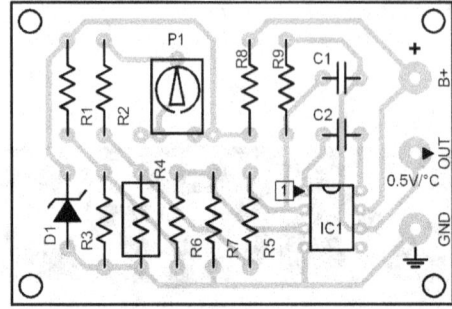

Figure 96.1 Parts Placement

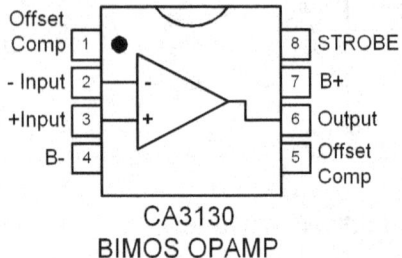

CA3130
BIMOS OPAMP

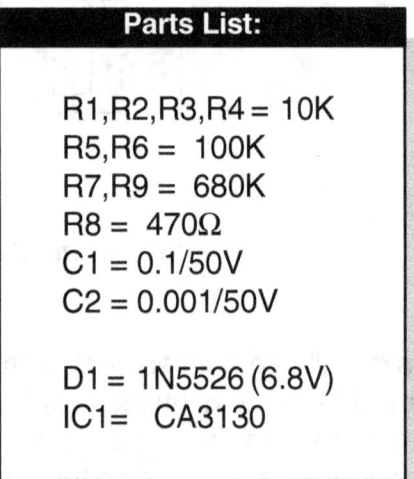

Parts List:
R1,R2,R3,R4 = 10K
R5,R6 = 100K
R7,R9 = 680K
R8 = 470Ω
C1 = 0.1/50V
C2 = 0.001/50V
D1 = 1N5526 (6.8V)
IC1= CA3130

97 KEY FINDER

This small circuit offers relief for those who always forget where they have placed their keys. One only needs to whistle with a tone frequency between 3 and 4 kHz (learning to whistle with this frequency is easy), and the key will answer with a beeping tone. Obviously, the keys must be attached to this circuit. The circuit can be constructed in such a size (much smaller than the one shown in figure 97.0) that it can be used as a key holder.

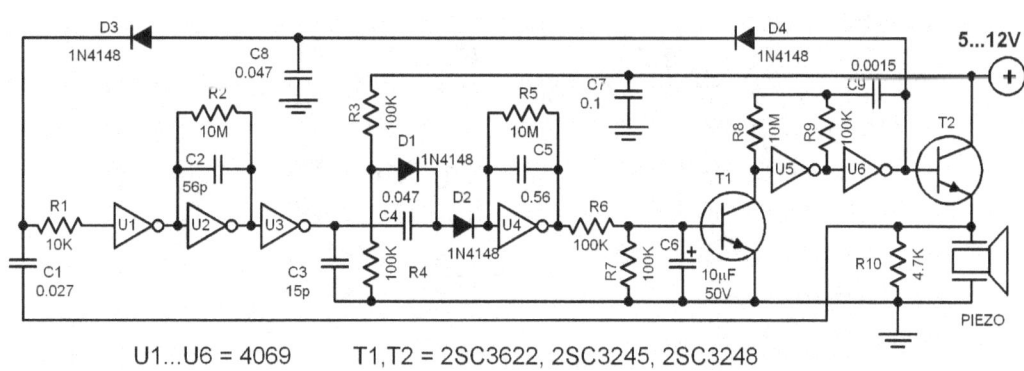

U1...U6 = 4069   T1,T2 = 2SC3622, 2SC3245, 2SC3248

Diagram 97.0   Key Finder

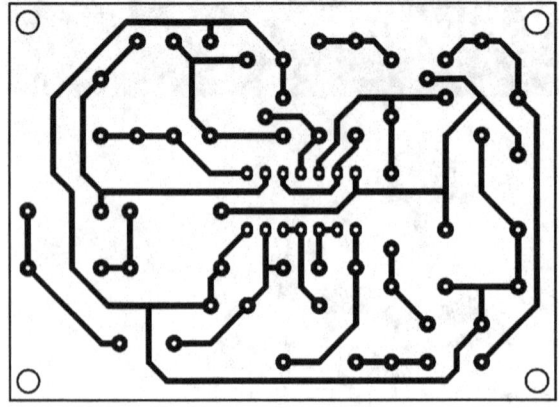

Figure 97.0 Printed Circuit Layout

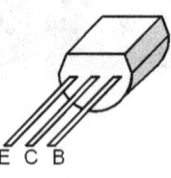

2SC3622
2SC3245
2SC3249

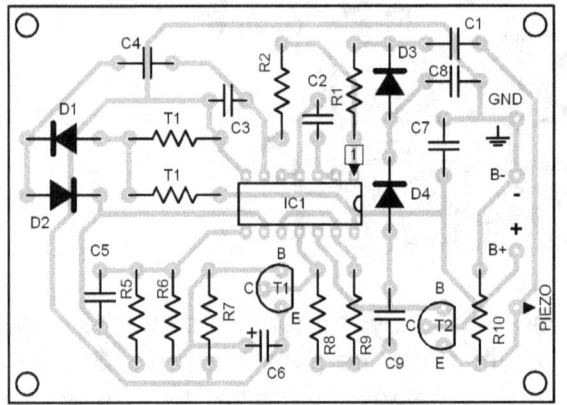

Figure 97.1 Parts Placement

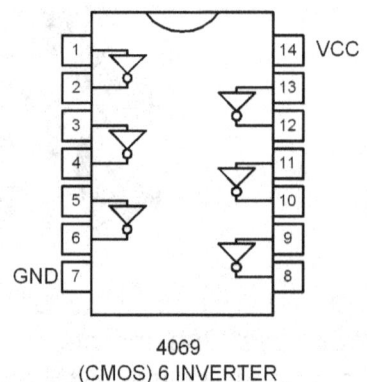

4069
(CMOS) 6 INVERTER

Parts List:	
R1 = 10K	C6 = 10µF/25V
R2,R5,R8 = 10M	C7 = 0.1/50V
R3,R4,R6,R7,R9 = 100K	C9 = 0.0015/50V
R10 = 4.7K	
C1 = 0.027/50V	D1,D2,D3,D4= 1N4148
C2 = 56p/50V	T1,T2 = 2SC3622
C3 = 15p/50V	(2SC3245)(2SC3248)
C4,C8 = 0.047/50V	IC1= 4069
C5 = 0.56/50V	

98 POWER ZENER

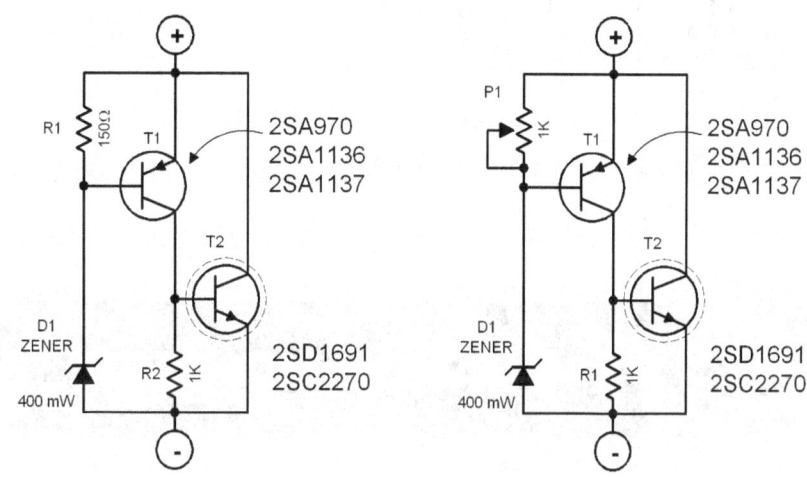

Diagram 98.0 Power Zener

The current handling capacity of a zener diode can be increased by adding a few more components to it as you can see in the circuit. This circuit can handle currents up to 500 mA at 25V. The zener voltage is however increased by 0.7V due to the transistors used. Potentiometer P1 offers flexibility. You can adjust the zener voltage within a limited range through P1. Afterwards, it can be replaced with a fixed value resistor. The transistor T2 must be heatsinked.

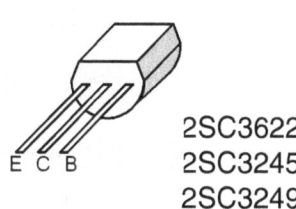

2SC3622
2SC3245
2SC3249

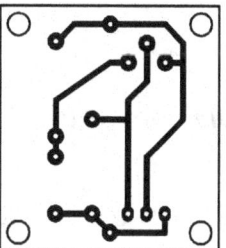

Figure 98.0
Printed Circuit Layout

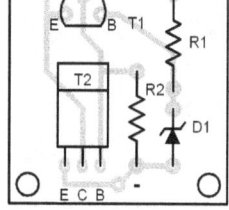

Figure 98.1
Parts Placement

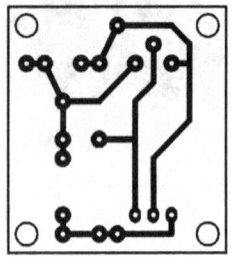

Figure 98.2
Printed Circuit Layout

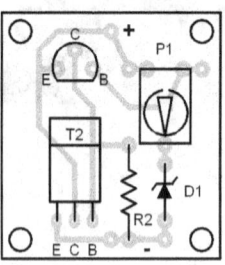

Figure 98.3
Parts Placement

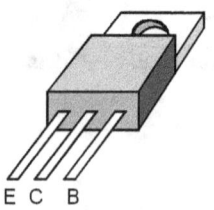

2SC2270
2SD1691

99   CHARGE MONITOR

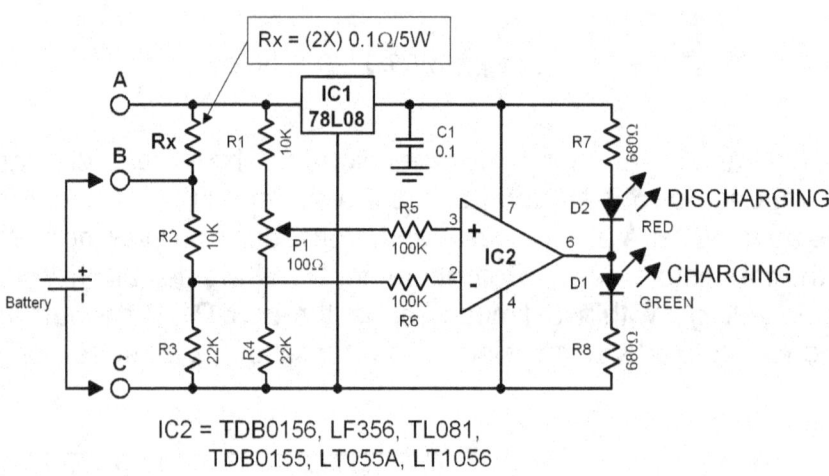

IC2 = TDB0156, LF356, TL081,
TDB0155, LT055A, LT1056

Diagram 99.0   Voltage Monitor

This circuit indicates the charging current flowing in a storage battery. The current being monitored is converted to a voltage value by Rx. Resistor Rx can be constructed from either one 0.1ohm/5W resistor or a fuse. The direction of the current flowing through this resistor is detected by the IC2. The circuit then displays through D1 or D2 whether the battery is being charged or discharged.

The display threshold can be adjusted slightly through P1. When D1 lights, the battery is being charged. Conversely when D2 lights, the battery is discharging. This circuit is designed for 12 volt batteries only.

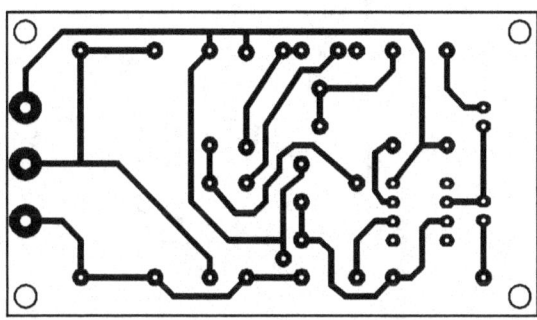

Figure 99.0 Printed Circuit Layout

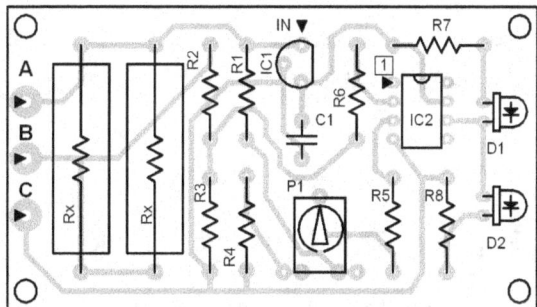

Figure 99.1 Parts Placement

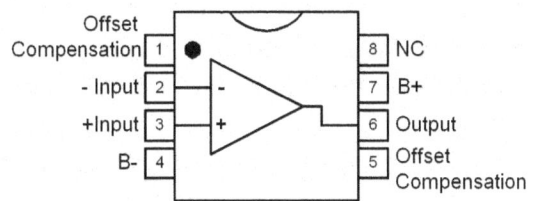

TDB0156, TDB0155, LF356
TL081, LT055A, LT1056

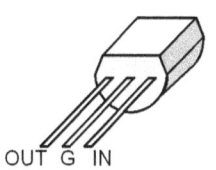

78L05

100 TEMPERATURE MONITOR

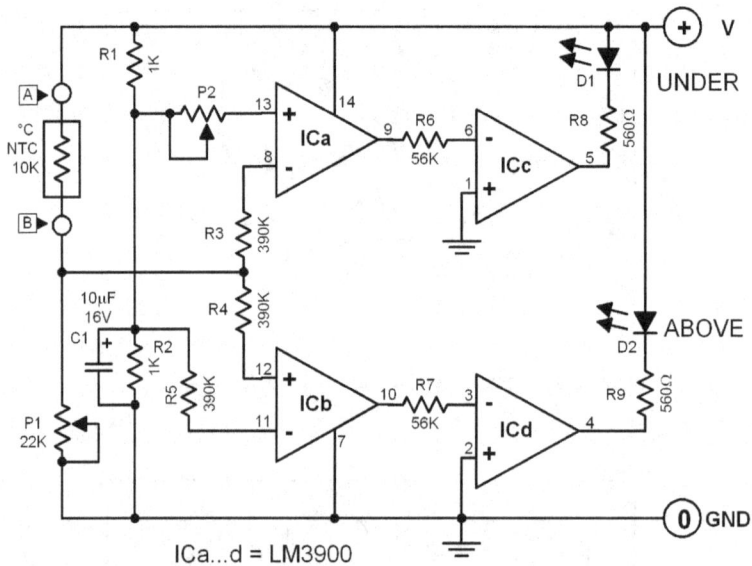

Diagram 100.0 Temperature Monitor

Sometimes a continuous analog display of a temperature value being monitored is not necessary. A simple indication whether the temperature value exceeded a maximum level or went below the minimum level is sometimes enough. The circuit here does just that. It indicates the temperature level above +25°C or below +20°C by lighting one of the two LEDs. The temperature sensor is an NTC resistor coupled to two comparators. When D3 lights, the temperature is above +25°C. Conversely when D2 lights, the temperature is below +20°C.

Calibration: Place the NTC resistor in cold water. Slowly heat the water until the desired maximum temperature level is reached (use a thermometer), then adjust P1 until D2 lights. To set the minimum level, stop heating the water then slowly add cold water to it until the desired lower temperture level is reached as indicated by the thermometer. This time, adjust P2 until D1 lights.

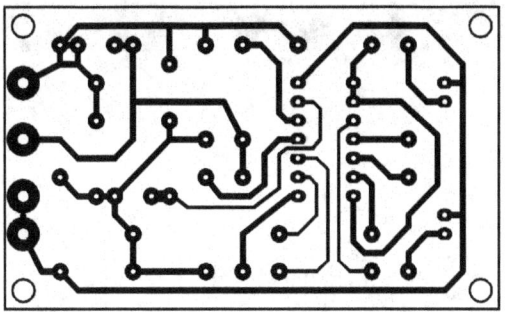

Figure 100.0 Printed Circuit Layout

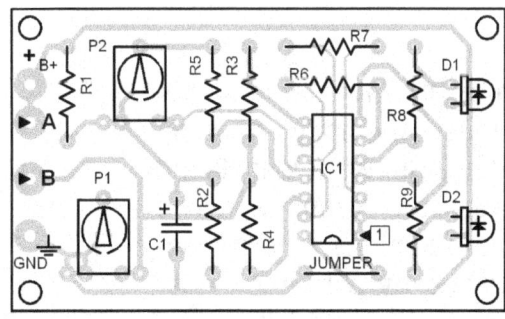

Figure 100.1 Parts Placement

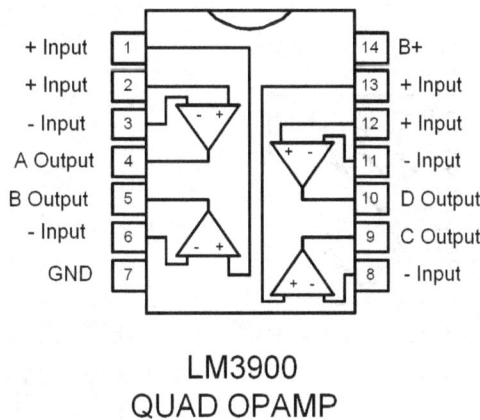

LM3900
QUAD OPAMP

101 MOS SIGNAL INJECTOR

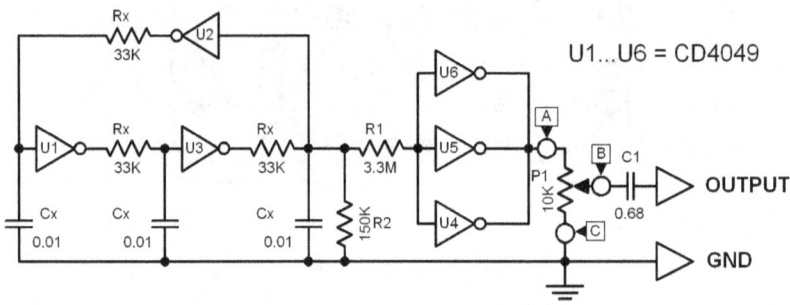

Diagram 101.0 MOS Signal Injector

A signal injector is a highly invaluable helper in troubleshooting jobs, specially in audio circuits. It can speed up your work in isolating the malfunctioning stage and component. The featured signal injector functions as a phase-shifted oscillator followed by a buffer stage. The output signal's frequency is dependent on the values of Rx and Cx. It can be found by using the formula: $f = 1/3.3(RxCx)$. The values given in the diagram cause the oscillator to generate a 1 kHz signal. Power supply can be between 3.5 and 14 volts. P1 varies the output's amplitude.

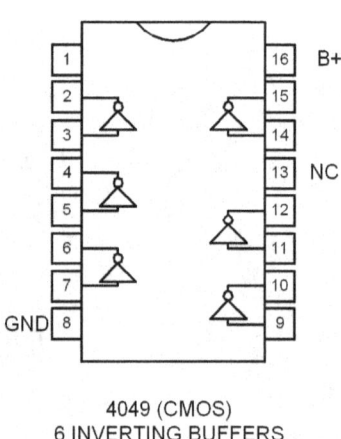

4049 (CMOS)
6 INVERTING BUFFERS

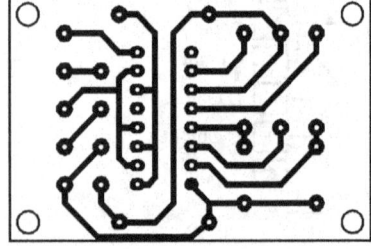

Figure 101.0
Printed Circuit Layout

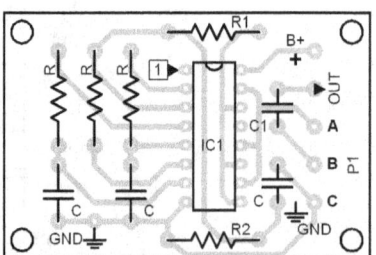

Figure 101.1
Parts Placement

APPENDICES

Specifications of the transistors used in the projects

Descriptive Part of the Table:

Type

The original type designation has been taken over directly from the manufacturers, with the abbreviation of the manufacturer added in brackets only in those cases in which different manufacturers used the same type designation.

Mat.

The materials used are abbreviated as follows:

Ge   Germanium
Mos   MOS technology (metal oxide silicon)
Si   Silicon
V-MOS   Vertical MOS technology

Pol.

The polarities used are abbreviated as follows:

npn   NPN structure
n-ch   N channel type (FET)
n-p   More than one transistor with different polarities in one case
pnp   PNP structure
p-ch   P channel type (FET)

Abbreviations used in the following table:

A	Antenna amplifer	**FET**	Field-effect transistor
AGC	Regulating steps	**FET-depl.**	Field-effecttransistor, depletion type
AF	AF range	**FET-enh.**	Field-effect transistor, enhancement type
AM	AM range		
CATV	Broad band cable amplifier	**FM**	FM range
CB	CB-radio	**fs**	Fast switch
CTV	Colour television application	**HD**	Horizontal deflection
chop	Chopper	**hi-rel**	high reliability
Darl	Darlington transistor	**Idss**	Drain source short-circuit current (FET)
dg	Dual Gate (FET)	**IF**	IF applications
double	Paired types	**in**	Input stages
dr	Driver stages	**iso**	insulated
dual	Dual transistor (differential amplifier)	**ln**	Low noise
		min	Miniaturised version
end	Final stages	**mix**	Mixer stages
		nixie	Digital display tube

osc	Oscillator stages
pow	Power stages
radiation	Aerospace applications (radiation-proof)
RF	RF range
s	Switch
SMP	Switch-mode power supply
SSB	Single side-band operation
Stabi	Stabilisation
sym	Symmetrical types
TV	Television applications

Ugs	Gate source voltage
UHF	UHF range > 250MHz
uni	Universal type
Up	Pinch-off voltage (FET)
VD	Vertical deflection
VHF	VHF range 100-250 MHz
Vid	Video output stages
+Diode, +di	With integrated diode
../..ns	turn-on/turn-off time

DATA PART

In the case of the ratings, either average values are quoted (< = max.) or lower (> = min.) guaranteed values. As a rule apply at 25°C, unless otherwise indicated.

Uc
With transistors, the usual situation is for U_{CBO}(colletor base reverse bias) to be quoted, or U_{CEO} and U_{CEO} (collector emitter reverse bias). With FETs, U_{DS} (drain source voltage) is always quoted.

Ic
With transistors, I_c (collector current) is always quoted. If this is followed by (ss) in brackets, I_{CM} is quoted, i.e. the peak value of the collector current. With FETs, I_D (drain current) is always quoted.

Ptot
As a rule, the total leakage power Ptot is quoted, with RF types we always quote the RF output power P_Q, with corresponding frequency in brackets.

Amplification
The DC current gain B(h_{FE}) or the short-circuit current gain ß(h_{fe}) are always quoted as guaranteed values.

fт
The transition frequency is always qouted in MHz.

Specifications of the transistors used in the projects

Type	Mat.	Pol.	Description	UC [Vmax]	IC [Amax]	Ptot [Wmax]	Current Gain	fT [MHz]
MJ3001	Si	npn	Darl+diode,pow	60	10.00	150.00($25°C)	>10	
MJE243	Si	npn	AF-s-pow	100	4.00	1.50($25°C)	40.120	>40.00
MJE244	Si	npn	AF-s-pow	100	4.00	1.50($25°C)	>25	>40.00
MJE253	Si	npn	AF-s-pow	100	4.00	1.50($25°C)	40-120	>40.00
MJE4350	Si	pnp	AF-end,s-pow	100	16.00	125.00($25°C)	15	>1.00
MJE5170	Si	pnp	uni-pow	120	6.00	2.00($25°C)	15-100	>1.00
MJE5180	Si	npn	uni-pow	120	6.00	2.00($25°C)	15-100	>1.00
MPF102	Si	n-ch	FET,VHF-in,sym,mix 25V,Idss>2mA,Up<V					
MPF106	Si	n-ch	FET,VHF 25V,Idss>4mA,Up<8V					
MPS-A29	Si	npn	Darl	100	0.50	1.50($25°C)	>10	>125.00
2N708	Si	npn	s	40/15	0.20	1.20(25°C)	>15	480.00
2N1711	Si	npn	uni	75	0.50	3.00(25°C)	75	>70.00
2N1889	Si	npn	AF-s	100/60	0.50	3.00(25°C)	40-120	>50.00
2N1890	Si	npn	AF-s	100/60	0.50	3.00(25°C)	100-300	>60.00
2N1990	Si	npn	nixie	100	1.00	2.00(25°C)	>25	
2N2102	Si	npn	AF-s	120/65	1.00	5.00(25°C)	40-120	>120.00
2N2222	Si	npn	ini	0		1.80(25°C)		
2N2368	Si	npn	fs	40/15	0.20	1.20(25°C)	20-60	>400.00
2N2369	Si	npn	fs	40/15	0.20	1.20(25°C)	40-120	>500.00
2N2905	Si	pnp	uni	60/40	0.60	3.00(25°C)	100-300	>200.00
2N2904	Si	pnp	uni	60/40	0.60	3.00(25°C)	40-120	>200.00
2N3019	Si	npn	uni	140/80	1.00	5.00(25°C)	100-300	>100.00
2N3020	Si	npn	uni	140/80	1.00	5.00(25°C)	40-120	>80.00
2N3055	Si	npn	AF-s-pow	100/60	15.00	115.00($25°C)	20-70	>2.50
2N3109	Si	npn	AF-s	80/40	1.00	5.00(25°C)	100-300	>70.00
2N3110	Si	npn	AF-s	80/40	1.00	5.00(25°C)	40-120	>60.00
2N3367	Si	n-ch	FET,uni,ln	40V,Idss>0.5mA,Up<2.5V				
2N3370	Si	n-ch	FET,uni,ln	40V,Idss>0.1mA,Up3.2V				
2N3454	Si	n-ch	FET,uni	50V,Idss>0.05mA,Up<2.3V				
2N3819	Si	n-ch	FET,VHF,uni,sym	25V,Idss>2mA,Up<8V				
2N3823	Si	n-ch	FET,VHF,ln	30V,Idss>4mA,Up<8V				
2N3903	Si	npn	uni	60/40	0.20	1.50(25°C)	50-150	>250.00
2N3904	Si	npn	uni	60/40	0.20	1.50(25°C)	100-300	>300.00
2N3905	Si	pnp	uni	40	0.20	1.50(25°C)	50-150	>200.0
2N3906	Si	pnp	uni	40	0.20	1.50(25°C)	100-300	>250.00
2N4118	Si	n-ch	FET,uni	40V,Idss>0.08mA,Up<3V				
2N5294	Si	npn	AF-s-pow	80/70	4.00	1.80($25°C)	30-120	>0.80
2N5397	Si	n-ch	FET,VHF/UHF	25V,Idss>10mA,Up<6V				
2N5398	Si	n-ch	FET,VHF/UHF	25V,Idss>5mA,Up<6V				
2N5486	Si	n-ch	FET,VHF/UHF	25V,Idss>8mA,Up<6V				
2N6038	Si	npn	Darl+diode,pow	60	4.00	1.50($25°C)	>10	>25.00
2N6039	Si	npn	Darl+diode,pow	80	4.00	1.50($25°C)	>10	>25.00
2N6283	Si	npn	Darl+diode,pow	80	20.00	160.00($25°C)	>10	>4.00
2N6284	Si	npn	Darl+diode,pow	100	20.00	160.0($25°C)	>10	>4.00
2N6412	Si	npn	AF-s-pow	60/40	4.00	15.00($25°C)	>5	>50.00
2N6414	Si	pnp	AF-s-pow	80/60	4.00	15.00($25°C)	>5	>50.00
2SA511	Si	pnp	AF/RF/s	90/80	1.50	8.00(25°)	30-150	60.00
2SA597	Si	pnp	RF-s	50/40	1.00	6.00($25°C)	10-250	400.00
2SA761	Si	pnp	uni	110	2.00	6.30($25°)	50-240	80.00
2SA970	Si	pnp	AF,ln	120	0.10	0.30(25°C)	200-700	100.0

Specifications of the transistors used in the projects

Type	Mat.	Pol.	Description	UC [Vmax]	IC [Amax]	Ptot [Wmax]	Current Gain	fT [MHz]
2SA1016	Si	pnp	uni,ln	120/100	0.05	0.40(25°)	160-960	110.00
2SA1123	Si	pnp	uni,ln	150	0.05	0.7(25°)	65-450	200.00
2SA1136	Si	pnp	AF-in,ln	120/100	0.10	0.30(25°C)	120-560	90.00
2SA1137	Si	pnp	AF-in,on	80	0.10	0.30(25°C)	120-560	90.00
2SA1141	Si	pnp	AF/Rf-pow	115	10.00	2.00($25°C)	100	80.00
2SA1285	Si	pnp	uni	120	0.20	0.90(25°C)	150-800	200.00
2SA1285A	Si	pnp	uni	150	0.10	0.90(25°C)	150-500	200.00
2SA1515	Si	pnp	uni	40/32	1.00	0.50(25°C)	82-390	150.00
2SA1705	Si	pnp	AF,s	60/50	1.00	0.90(25°C)	>30	150.00
2SA1706	Si	pnp	AF-s	60/50	2.00	1.00(25°C)	>40	150.00
2SB633	Si	pnp	AF-s-pow	100/85	6.00	40.00($25°C)	40-320	15.00
2SB764	Si	pnp	uni	60/50	1.00	0.90(25°C)	60-320	150.00
2SB822	Si	pnp	Af-dr/end	40/32	2.00	0.75(25°C)	82-390	100.00
2SB826	Si	pnp	s-pow	60/50	7.00	60.00($25°C)	>30	10.00
2SB867	Si	pnp	AF/s-pow,lo-sat	130/80	3.00	30.00($25°C)	60-260	30.00
2SB868	Si	pnp	AF/s-pow,lo-sat	130/80	4.00	35.00($25°C)	60-260	30.00
2SB869	Si	pnp	AF/s-pow,lo-sat	130/80	5.00	40.00($25°C)	60-260	30.00
2SB870	Si	pnp	AF/s-pow,lo-sat	120/80	7.00	40.00($25°C)	60-260	30.00
2SB874	Si	pnp	AF/s-pow, TV-VD	100/60	2.00	20.00($25°C)	>40	250.00
2SB909	Si	pnp	AF-dr/end	40/32	1.00	1.00(25°C)	82-390	150.00
2SB911	Si	pnp	AF-dr/end	40/32	2.00	1.00(25°C)	82-390	100.0
2SB920	Si	pnp		120/80				
2SB921	Si	pnp		120/80				
2SB1064	Si	pnp	AF-s-pow	60/50	3.00	1.50($25°)	60-320	70.00
2SB1114	Si	pnp	min,uni	20	2.00	2.00($25°C)	135-600	180.00
2SB1116	Si	pnp	uni	60/50	1.00	0.75(25°C)	135-600	120.00
2SB1142	Si	pnp	s-pow	60/50	2.50	10.00(25°C)	>35	140.00
2SB1143	Si	pnp	s-pow	60/50	4.00	10.00(25°C)	>40	150.00
2SB1144	Si	pnp	AF/s-pow,lo-sat	120/100	1.50	10.00(25°C)	>30	100.00
2SB1230	Si	pnp	AF/s-pow,lo-sat	110/100	15.00	100.00($25°C)	50-140	
2SB1231	Si	pnp	AF/s-pow,lo-sat	110/100	25.00	120.00($25°C)	50-140	
2SB1232	Si	pnp	AF/s-pow,lo-sat	110/100	40.00	150.00($25°C)	50-140	
2SC270	Si	npn	s-pow	270/75	5.00	50.00($25°C)	24-40	22.00
2SC460	Si	npn	AM-in/mix/osc	30	0.10	0.20(25°C)	35-200	230.00
2SC696	Si	npn	uni	100/60	3.00	0.75(25°C)	30-173	100.00
2SC763	Si	npn	VHF	25/12	0.02	0.10(25°C)	20-300	>400.00
2SC829	Si	npn	AM/FM-in/mix/osc	30/20	0.03	0.40(25°C)	40-500	230.00
2SC959	Si	npn	uni	120/80	0.70	0.70(25°C)	40-200	100.00
2SC1324	Si	npn	UHF-CATV	35/25	0.15	3.00(25°C)	10-35	
2SC1876	Si	npn	Darl	100/70	0.50	0.80(25°C)	>20	
2SC2124	Si	npn	TV-HD	220/800	2.00	5.00($90°C)	20	4.00
2SC2125	Si	npn	TV-HD	220/800	5.00	50.00($25°C)	8-25	5.00
2SC2270	Si	npn	lo-sat	50/20	5.00	1.00($25°C)	>70	100.00
2SC2334	Si	npn	s-pow,dc-dc conv.	150/100	7.00	40.00($25°C)	>20	
2SC2459	Si	npn	uni	120	0.10	0.20(25°C)	200-700	100.00
2SC2675	Si	npn	AF,ln	80	0.10	0.30(25°C)	180-820	120.00
2SC2724	Si	npn	FM-IF	30/25	0.03	0.20(25°C)	25-300	200.00
2SC3112	Si	npn	AF,ln	50	0.15	0.40(25°C)	600-3600	250.00
2SC3179	Si	npn	AF-pow	80/60	4.00	30.00($25°C)	100	15.00
2SC3245	Si	npn	uni	120	0.10	0.90(25°C)	150-800	200.00

Specifications of the transistors used in the projects

Type	Mat.	Pol.	Description	UC [Vmax]	IC [Amax]	Ptot [Wmax]	Current Gain	fT [MHz]
2SC3245A	Si	npn	uni	150	0.10	0.90(25°C)	400-800	200.00
2SC3248	Si	npn	uni	180	0.10	0.90(25°C)	150	130.00
2SC3358	Si	npn	UHF	20/12	0.10	0.25(25°C)	50-300	7000.00
2SC3420	Si	npn	lo-sat	50/20	5.00	10.00(25°C)	>70	100.00
2SC3622	Si	npn	AF-s,hi-beta	60/50	0.15	0.25(25°C)	1000-3200	250.00
2SC4308	Si	npn	VHF-A	30/20	0.30	0.60(25°C)	50-200	2500.00
2SD386	Si	npn	TV-VD	200/120	3.00	1.75($25°C)	40-320	8.00
2SD406	Si	npn	Darl	100	2.00	15.00(25°C)	>2000	
2SD613	Si	npn	AF-s-pow	100/85	6.00	40.00($25°C)	40-320	15.00
2SD614	Si	npn	Darl	100/80	3.00	0.80(25°C)	3000	15.00
2SD621	Si	npn	TV_HD	2500/900	3.00	50.00($25°C)	3-15	
2SD628	Si	npn	Darl+diode,pow	100	10.00	80.00($25°C)	>1000	
2SD629	Si	npn	Darl+diode,pow	100	10.00	100.00($25°C)	>1000	
2SD688	Si	npn	Darl,pow	100	1.50	0.80($25°C)	>10	
2SD712	Si	npn	AF-s-pow	100	4.00	30.00($25°C)	55-300	8.00
2SD726	Si	npn	AF-s-pow	100/80	4.00	40.00($25°C)	35-320	10.00
2SD729	Si	npn	Darl+diode,pow	100	20.00	125.00($25°C)	>1000	
2SD781	Si	npn	s-pow,TV-HD	150/60	2.00	1.00(25°C)	150	
2SD826	Si	npn		60/20	5.00	1.00($25°C)	120-560	120.00
2SD838	Si	npn	TV-HD,s-pow	2500/900	3.00	50.00($25°C)	3-15	
2SD892A	Si	npn	Darl	60/50	0.50	0.40(25°C)	>8000	150.00
2SD1049	Si	npn	AF-s-pow	120/80	25.00	80.00($25°C)	>20	
2SD1062	Si	npn	s-pow	60/50	12.00	40.00($25°C)	>30	10.00
2SD1153	Si	npn	Darl	80750	1.50	0.90(25°C)	>40	120.00
2SD1177	Si	npn	AF-pow,TV-HD	100/60	2.00	20.00($25°C)	>40	230.00
2SD1237	Si	npn	s-pow	90/80	7.00	1.75($25°C)	>30	20.00
2SD1238	Si	npn	s-pow	90/80	12.00	80.00($25°C)	>30	20.00
2SD1639	Si	npn	AF-s-pow	100/80	2.20	10.00($25°C)	40-200	
2SD1684	Si	npn	AF/s-pow,lo-sat	120/100	1.50	10.00(25°C)	>30	120.00
2SD1685	Si	npn	AF/s-pow,lo-sat	60/20	5.00	10.00(25°C)	>95	120.00
2SD1691	Si	npn	AF-s-pow	60	5.0	20.00(25°C)	100-400	
2SD1840	Si	npn	AF/s-pow,lo-sat	110/100	15.00	100.00($25°C)	50-140	
2SD1841	Si	npn	AF/s-pow,lo-sat	110/100	25.00	120.00($25°C)	50-140	
2SD1842	Si	npn	AF/s-pow,lo-sat	110/100	40.00	150.00($25°C)	50-140	
2SD2116	Si	npn	Darl	80/50	0.70	1.00(25°C)	>40	
2SD2117	Si	npn	Darl	80/50	1.50	1.00(25°C)	>30	
2SD2213	Si	npn	Darl,AF	150/80	1.50	0.90(25°C)	>10	
2SJ165	V-MOS	p-ch	FET-enh.,	50V,0.1A,0.25W				
2SK422	V-MOS	n-ch	FET-enh.	60v,0.7A,0.9W,17/12ns				
2SK423	V_MOS	n-ch	FET-enh.	100V,0.5A,0.9W,15/20ns				
3N140	MOS	n-ch	FET-depl.,dg,FM/VHF-in	20V,Idss>5mA				
3N225	MOS	n-ch	FET-depl.,dg,UHF	25V,Idss>1mA,Up<4V				
3SK35	MOS	n-ch	FET-depl.,dg,VHF	20V,Idss>3mA,Up<4V				
3SK37	MOS	n-ch	FET-depl.,dg,VHF	20V,Idss>4mA,Up<3V				
3SK45	MOS	n-ch	FET-depl.,dg,VHF	22V,Idss>4mA,Up<3V				
3SK61	MOS	n-ch	FET-depl.,dg,VHF	20V,Idss>4mA,Up<3V				
3SK72	MOS	n-ch	FET-depl.,dg,VHF	20V,Idss>2.5mA,Up<3V				
3SK77	MOS	n-ch	FET-depl.,dg,VHF	20V,Idss>3mA,Up<2.5V				
3SK85	MOS	n-ch	FET-depl.,dg,VHF	20V,Idss>4mA,Up<3V				

SEMICONDUCTOR DIODE SPECIFICATIONS

Device	Type	Material	Peak Inverse Voltage, PIV (Volts)	Average Rectified Current Forward (Reverse) IO (A) (IR(A))	Peak Surge Current, IFSM 1 sec. @ 25ºC (A)	Average Forward Voltage, VF (Volts)
1N34	Signal	Germanium	60	8.5 m (15.0m)		1.0
1N34A	Signal	Germanium	60	5.0 m (30.0m)		1.0
1N67A	Signal	Germanium	100	4.0 m (5.0m)		1.0
1N191	Signal	Germanium	90	5.0 m	1.0	
1N270	Signal	Germanium	80	0.2 (100 m)		1.0
1N914	Fast Switch	Silicon (Si)	75	75.0 m (25.0 n)	0.5	1.0
1N1184	RFR	Si	100	35 (10 m)		1.7
1N2071	RFR	Si	600	0.75 (10.0m)		0.6
1N3666	Signal	Germanium	80	0.2 (25.0m)		1.0
1N4001	RFR	Si	50	1.0 (0.03 m)		1.1
1N4002	RFR	Si	100	1.0 (0.03 m)		
1N4003	RFR	Si	200	1.0 (0.03 m)		1.1
1N4004	RFR	Si	400	1.0 [0.03 m]		1.1
1N4005	RFR	Si	600	1.0 (0.03 m)		1.1
1N4006	RFR	Si	800	1.0 (0.03 m)		1.1
1N4007	RFR	Si	1000	1.0 (0.03 m)		1.1
1N4148	Signal	Si	75	10.0 m (25.0 n)		1.0
1N4149	Signal	Si	75	10.0 m (25.0 n)		1.0
1N4152	Fast Switch	Si	40	20.0 m (0.05m)		0.8
1N4445	Signal	Si	100	0.1 (50.0 n)		1.0
1N5400	RFR	Si	50	3.0	200	
1N5401	RFR	Si	100	3.0	200	
1N5402	RFR	Si	200	3.0	200	
1N5403	RFR	Si	300	3.0	200	
1N5404	RFR	Si	400	3.0	200	
1N5405	RFR	Si	500	3.0	200	
1N5406	RFR	Si	600	3.0	200	
1N5767	Signal	Si		0.1 (1.0μ)		1.0
ECG5863	RFR	Si	600	6	150	0.9

* RFR = Rectifier, Fast Recovery

ZENER DIODES SPECIFICATIONS

Zener Voltage (Volts)	Power (Watts)							
	0.25	0.4	0.5	1.0	1.5	5.0	10.0	50.0
1.8	1N4614							
2.0	1N4615							
2.2	1N4616							
2.4	1N4617	1N4370,A	1N4370,A,1N5221,B 1N5985,B					
2.5			1N5222B					
2.6	1N702,A							
2.7	1N4618	1N4371,A	1N4371,A,1N5223,B 1N5839, 1N5986					
2.8			1N5224B					
3.0	1N4619	1N4372,A	1N4372,1N5225,B 1N5987					
3.3	1N4620	1N746,A 1N764 A 1N5518	1N746A 1N5226,B 1N5988	1N3821 1N4728,A	1N5913	1N5333,B		
3.6	1N4621	1N747,A 1N5519	1N747A 1N5227,B,1N5989	1N3822 1N4729,A	1N5914	1N5334,B		
3.9	1N4622	1N748,A 1N5520	1N748A,1N5228,B 1N5844, 1N5990	1N3823 1N4730,A	1N5915	1N5335,B	1N3993A	1N4549,B 1N4557,B
4.1	1N704,A							
4.3	1N4623	1N749,A 1N5521	1N749,A 1N5229,B 1N5845,1N5991	1N3824 1N4731 ,A	1N5916	1N5336,B	1N3994,A	1N4550,B 1N4558,B
4.7	1N4624	1N750,A 1N5522	1N750A ,1N5230,B 1N5846, 1N5992	1N3825 1N4732,A	1N5917	1N5337,B	1N3995,A	1N4551,B 1N4559,B
5.1	1N4625 1N4689	1N751 A 1N5523	1N751A, 1N5231,B 1N5847,1N5993	1N3826 1N4733	1N5918	1N5338,B 1N4560,B	1N3996,A	1N4552,B
5.6	1N708A 1N4626	1N752,A 1N5524	1N752,A,1N5232,B 1N5848, 1N5994	1N3827 1N4734,A	1N5919	1N5339,B 1N4561,B	1N3997,A	1N4553,B
5.8	1N706A	1N762						
6.0				1N5233B 1N5849			1N5340,B	
6.2	1N709,1N4627 MZ605,MZ610 MZ620,MZ640	1N753,A 1N821,3,5, 7,9; A	1N753,A 1N5234,B, 1N5850 1N5995	1N3828,A 1N4735,A	1N5920	1N5341,B 1N4562,B	1N3998,A	1N4554,B
6.4	1N4565-84,A							
6.8	1N4099	1N754,A 1N957,B 1N5526	1N754,A 1N757,B 1N5235,B 1N5851 1N5996	1N3016,B 1N3829 1N4736,A	1N3785 1N5921	1N5342,B	1N2970,B 1N3999,A	1N2804B 1N3305B 1N4555, 1N4563
7.5	1N4100	1N755,A 1N958,B 1N5527	1N755A,1N958,B 1N5236,B, 1N5862 1N5997	1N3017,A,B 1N3830 1N4737,A	1N3786 1N5922	1N5343,B 1N4000,A 1N4556,	1N2971,B 1N3306,B	1N2805,B 1N4564
8.0	1N707A							
8.2	1N712A 1N4101	1N756,A 1N959,B 1N5528	1N756,A 1N959,B,1N5237,B 1N5853 ,1N5998	1N3018,B 1N4738,A	1N3787 1N5923	1N5344,B	1N2972,B	1N2806,B 1N3307,B
8.4		1N3154-57,A 1N3155-57	1N3154,A					
8.5	1N4775-84,A		1N5238,B,1N5854					
8.7	1N4102					1N5345,B		
8.8		1N 764						
9.0		1N764A	1N935-9;A,B					

ZENER DIODES SPECIFICATIONS

Zener Voltage (Volts)	Power (Watts)							
	0.25	0.4	0.5	1.0	1.5	5.0	10.0	50.0
9.1	1N4103	1N757,A 1N960,B 1N5529	1N757,A, 1N960,B 1N5239,B, 1N5855 1N5999	1N3019,B 1N4739,A	1N3788 1N5924	1N5346,B	1N2973,B	1N2807,B 1N3308,B
10.0	1N4104	1N758,A 1N961,B 1N5530,B	1N758,A, 1N961,B 1N5240,B, 1N5856 1N6000	1N3020,B 1N4740	1N3789 1N5925	1N5347,B	1N2974,B	1N2808,B 1N3309,A,B
11.0	1N715,A 1N4105	1N962,B 1N5531	1N962,B,1N5241,B 1N5857, 1N6001 1N941,A,B	1N3021,B 1N4741,A	1N3790 1N5926	1N5348,B	1N2975,B	1N2809,B 1N3310,B
11.7	1N716,A 1N4106							
12.0		1N759,A 1N963,B 1N5532	1N759,A ,1N963,B 1N5242,B, 1N5858 1N6002	1N3022,B 1N4742,A	1N3791 1N5927	1N5349,B	1N2976,B	1N2810,B 1N3311 ,B
13.0	1N4107	1N964,B 1N5533	1N964,B,1N5243,B 1N5859,1N6003	1N3023,B 1N4743,A	1N3792 1N5928	1N5350,B	1N2977,B	1N2811 ,B 1N3312,B
14.0	1N4108	1N5534	1N5244B, 1N5860			1N5351,B	1N2978,B	1N2812,B 1N3313,B
15.0	1N4109	1N965,B 1N5535	1N965,B,1N5245,B 1N5861,1N6004	1N3024,B 1N4744A	1N3793 1N5929	1N5352,B	1N2979,A,B	1N2813,A,B 1N3314,B
16.0	1N4110	1N966,B 1N553,B	1N966,B,1N5246,B 1N5862, 1N6005	1N3025,B 1N4745,A	1N3794 1N5930	1N5353,B	1N2980,B	1N2814,B 1N3315,B
17.0	1N4111	1N5537	1N5247,B 1N5863			1N5354,B	1N2981B	1N2815,B 1N3316,B
18.0	1N4112	1N967,B 1N5538	1N967,B 1N5248,B 1N5864, 1N6006	1N3026,B 1N4746,A	1N3795 1N5931	1N5355,B	1N2982,B	1N2816,B 1N3917,B
19.0	1N4113	1N5539	1N5249,B 1N5865			1N5356,B	1N2983,B	1N2817,B 1N3318,B
20.0	1N4114	1N968,B 1N5540	1N968,B,1N5250,B 1N5866, 1N6007	1N3027,B 1N4747,A	1N3796 1N5932,A,B	1N5357,B	1N2984,B	1N2818,B 1N3319,B
22.0	1N4115	1N959,B 1N5541	1N969,B,1N5241,B 1N5867, 1N6008	1N3028,B 1N4748,A	1N3797 1N5933	1N5358,B	1N2985,B	1N2819,B 1N3320,A,B
24.0	1N4116	1N5542 1N9701B	1N970,B,1N5252,B 1N586,1N6009	1N3029,B 1N4749,A	1N3798 1N5934	1N5359,B	1N2986,B	1N2820,B 1N3321 ,B
25.0	1N4117	1N5543	1N5253,B 1N5869			1N5360,B	1N2987B	1N2821,B 1N3322,B
27.0	1N4118	1N971,B	1N971,1N5254,B 1N5870,1N6010	1N3030,B 1N4750,A	1N3799 1N5935	1N5361,B	1N2988,B	1N2822B 1N3323,B
28.0	1N4119	1N5544	1N5255,B,1N5871			1N5362,B		
30.0	1N4120	1N972,B 1N5546	1N972,B,1N5256,B 1N5872,1N6011	1N3031,B 1N4751,A	1N3800 1N5936	1N5363,B	1N2989,B	1N2823,B 1N3324,B
33.0	1N4121	1N973,B 1N5546	1N973,B,1N5257,B 1N5873,1N6012	1N3032,B 1N4752,A	1N3801 1N5937	1N5364,B	1N2990,A,B	1N2824,B 1N3325,B
36.0	1N4122	1N974,B	1N974,B,1N5258,B 1N6874,1N6013	1N3033,B 1N4753,A	1N3802 1N5938	1N5365,B	1N2991,B	1N2825,B 1N3326,B
39.0	1N4123	1N975,B	1N975,B, 1N5259,B 1N5875 ,1N6014	1N3034,B 1N4754,A	1N3803 1N5939	1N5366,B	1N2992,B	1N2826,B 1N3327,B
43.0	1N4124	1N976,B	1N976,B,1N5260,B 1N5876,1N6015	1N3035,B 1N4755,A	1N3804 1N5940	1N5367,B	1N2993,A,B	1N2827,B 1N3328,B
45.0			1N2994B	1N2828B 1N3329B				

POWER FETs

Device No.	Type	Max. Diss. (W)	Max. VDS (Volts)	Max. ID (A)*	Gfs mmhos (typ.)	Input Ciss (pF)	Output Coss (pF)	Approx. Upper Freq. (MHz)	Case	Pack-Type Mnfr.	General applications age/
DV1202S	N-Chan.	10	50	0.5	100k	14	20	500	.380 SOE	1/S	RF power amp., oscillator
DV1202W	N-Chan.	10	50	0.5	100k	14	20	500	C-220	5/S	RF power amp., oscillator
DV1205S	N-Chan.	20	50	1	200k	26	38	500	.380 SOE	1/S	RF power amp., oscillator
DV1205W	N-Chan.	20	50	1	200k	26	98	500	C-220	5/S	RF power amp., oscillator
2SK133	N-Chan.	100	120	7	1M	600	350	1	TO-3	6/H	AF pwr. amp., switch (complem to 2SJ48)
2SK134	N-Chan.	100	140	7	1M	600	350	1	TO-3	6/H	AF pwr. amp., switch (complem to 2SJ49)
2SK135	N-Chan.	100	160	7	1M	600	350	1	TO-3	6/H	AF pwr. amp., switch (complem to 2SJ50)
2SJ48	P-Chan.	100	120	7	1M	900	400	1	TO-3	6/H	AF pwr. amp., switch (complem to 2SK133)
2SJ49	P-Chan.	100	140	7	1M	900	400	1	TO-3	6/H	AF pwr. amp., switch (complem to 2SK134)
2SJ50	P-chan.	100	160	7	1M	900	400	1	TO-3	6/H	AF pwr. amp., switch (complem to 2SK135)
VMP4	N-Chan.	25	60	2	170K	32	4.8	200	.380 SOE	1/S	VHF pwr. amp., rcvr front end (rf amp., mixer).
VN10KM	N-Chan.	1	60	0.5	100K	48	16	u	TO-92	2/S	High-speed line driver, relay driver, LED stroke driver
VN64GA	N-Chan.	80	60	12.5	150K	700	325	30	TO-3	3/S	Linear amp., power-supply switch, motor control
VN66AF	N-Chan.	15	60	2	150K	50	50	-	TO-202	4/S	High-speed switch, HF linear amp., audio amp. line driver.
VN66AK	N-Chan.	8.3	60	2	250K	93	6	100	TO-39	7/S	RF pwr. amp., high-current analog switching
VN67AJ	N-Chan.	25	60	2	250K	33	7	100	TO-3	3/S	RF pwr. amp., high-current switching
VN89AA	N-Chan.	25	80	2	250K	50	10	100	TO-3	3/S	High-speed switching, HF linear amps., line drivers.
IRF100	N-Chan.	125	80	16	300K	900	25	-	TO-3	3/S	High-speed switching, audio inverters.
IRF101	N-Chan.	125	60	16	300K	900	25	-	TO-3	3/S	Same as IRF100

Legend: * 25ºC (case)   S = M/A-COM   H = Hitachi   IR = International Rectifier.   Mnfr = Manufacturer

SMALL-SIGNAL FETs

Device No.	Type	Max. Diss. (mW)	Max. V_{DS} (Volts)	Max. I_D (mA)#	Min G_{fs} (mS)	Input C (pF)	$V_{GS(off)}$ (volts)	Upper Freq. (MHz)	Noise Figure	Case Type (typ)	/Mnfr.	General applications
2N4416	N-JFET	300	30	-15	4.5K	4	-6	450	400 MHz 4 dB	TO-72	1/S,M	VHF/UHF/RF amp.mix., osc.
2N5484	N-JFET	310	25	30	2.5K	5	-3	200	200 MHz 4 dB	TO-92	2/M	VHF/UHFamp.mix., osc.
2N5485	N-JFET	310	25	30	3.5K	5	-4	400	400 MHz 4 dB	TO-92	2/S	VHF/UHF/RF amp.mix., osc.
3N200	N-Dual-Gate MOSFET	330	20	50	10K	4-8.5	-6	500	400 MHz 4.5 dB	TO-72	3/R	VHF/UHF/RF amp.mix., osc.
3N202	N-Dual-Gate MOSFET	360	25	50	8K	6	-5	200	200 MHz 4.5 dB	TO-72	3/S	VHF amp., mixer
MPF102	N-JFET	310	25	20	2K	4.5	-8	200		TO-92	2/N,M	HF/VHF amp.,mix., osc.
MPF106/ 2N5484	N-JFET	310	25	30	2.5K	5	-6	400	200 MHz 4 dB	TO-92	2/N,M	HF/VHF/UHF amp.,mix.,osc.
40673	N-Dual-Gate MOSFET	330	20	50	12K	6	-4	400	200 MHz 6 dB	TO-72	3/R	HF/VHF/UHF amp. mix., osc.
U300	P-JFET	300	-40	20	8K	-50	+10	*	400 MHz	TO-18	4/S	General Purpose amp.
U304	P-JFET	350	-30	-50		27	+10	*	*	TO-18	4/S	analog switch, chopper
U310	N-JFET	500	30	60	10K	2.5	-6	450	450 MHz 3.2 dB	TO-52	5/S	common-gate VHF/UHF amp.,osc., mixer
U350	N-JFET Quad	1W	25	60	9K	5	-6	100	100 MHz 7 dB	TO-99	6/S	matched JFET doubly bal. mixer
U431	N-JFET Dual	300	25	30	10K	5	-6	100	*	TO-99	7/S	matched JFET cascade amp., balanced mixer

#25°C   S = Siliconix Inc.   R = RCA   N = National Semiconductor   M = Motorola

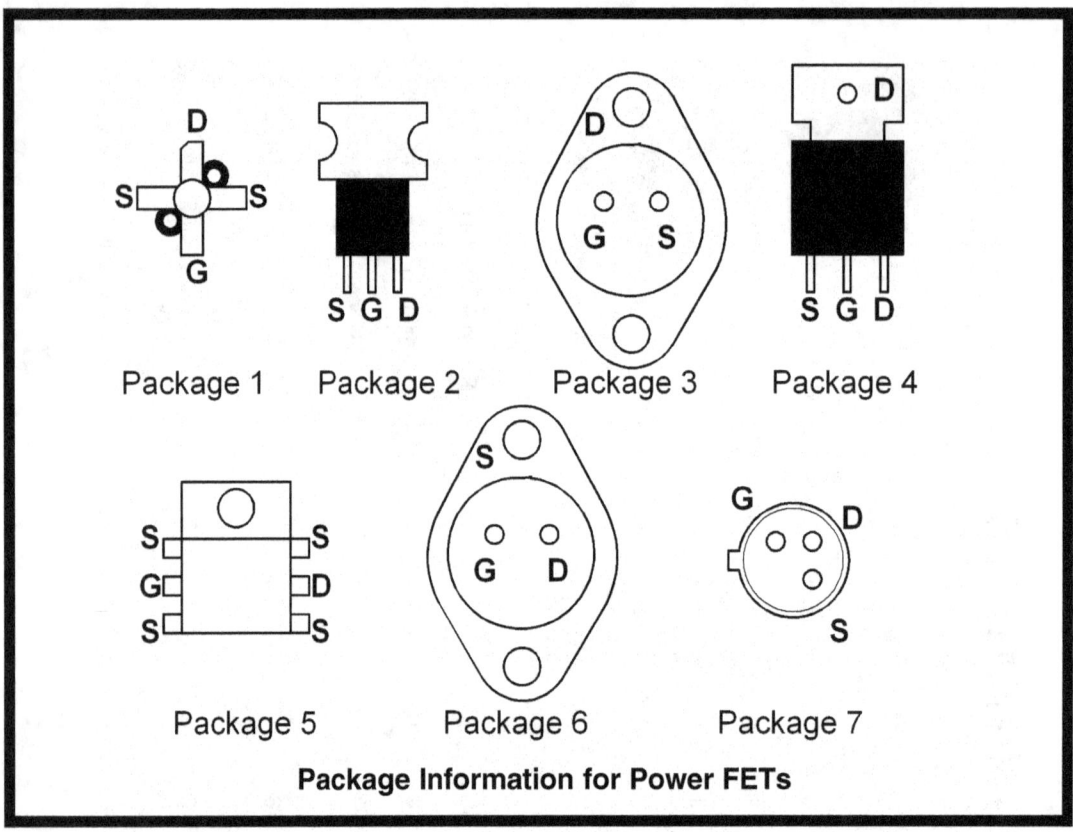

Package Information for Power FETs

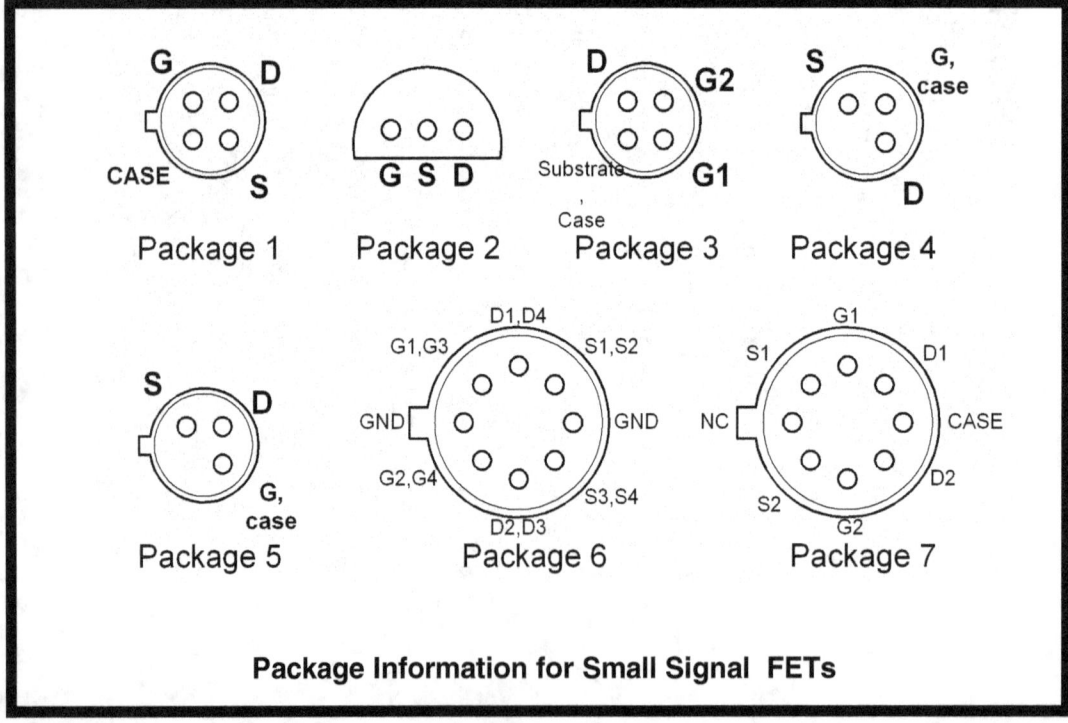

Package Information for Small Signal FETs

Three-Terminal Voltage Regulators

\# Listed numerically by device

Device	Description	Voltage	Current (Amps)	Package
317	Adj. Pos	+1.2 to +37	0.5	TO-205
317	Adj. Pos	+1.2 to +37	1.5	TO-204,TO-220
317L	Low Current Adj. Pos	+1.2 to +37	0.1	TO-205,TO-92
317M	Med Current Adj. Pos	+1.2 to +37	0.5	TO-220
350	High Current Adj. Pos	+1.2 to +33	3.0	TO-204,TO-220
337	Adj. Neg	-1.2 to -37	0.5	TO-205
337	Adj. Neg	-1.2 to -37	1.5	TO-204,TO-220
337M	Med Current Adj. Neg	-1.2 to -37	0.5	TO-220
309		+5	0.2	TO-205
309		+5	1.0	TO-204
323		+5	9.0	TO-204,TO-220
140-XX	Fixed Pos	**Note ***	1.0	TO-204,TO-220
340-XX			1.0	TO-204,TO-220
78XX			1.0	TO-204,TO-220
78LXX			0.1	TO-205,TO-92
78MXX			0.5	TO-220
78TXX			3.0	TO-204
79XX	Fixed Neg	**Note ***	1.0	TO-204,TO-220
79LXX			0.1	TO-205,TO-92
79MXX			0.5	TO-220

Legend:

Adj.	= Adjustable
Med	= Medium
Neg	= Negative
Pos	= Positive

Note * - XX indicates the regulated voltage; which may be anywhere from 1.2 volts to 35 volts. For example a 7808 is a positive 8-volt regulator, and a 7912 is a negative 12-volt regulator.

The regulator package may be denoted by an additional suffix, according to the following:

Package	Suffix
TO-204 (TO-3)	K
TO-220	T
TO-205 (TO-39)	H,G
TO-92	P,Z

Example:
A 7815K is a positive 15-volt regulator in a TO-204 package. An LM340T-8 is a positive 8-volt regulator in a TO-220 package. In addition, manufacturers use different prefixes. An LM7812 is equivalent to a µA 7812 or MC7812.

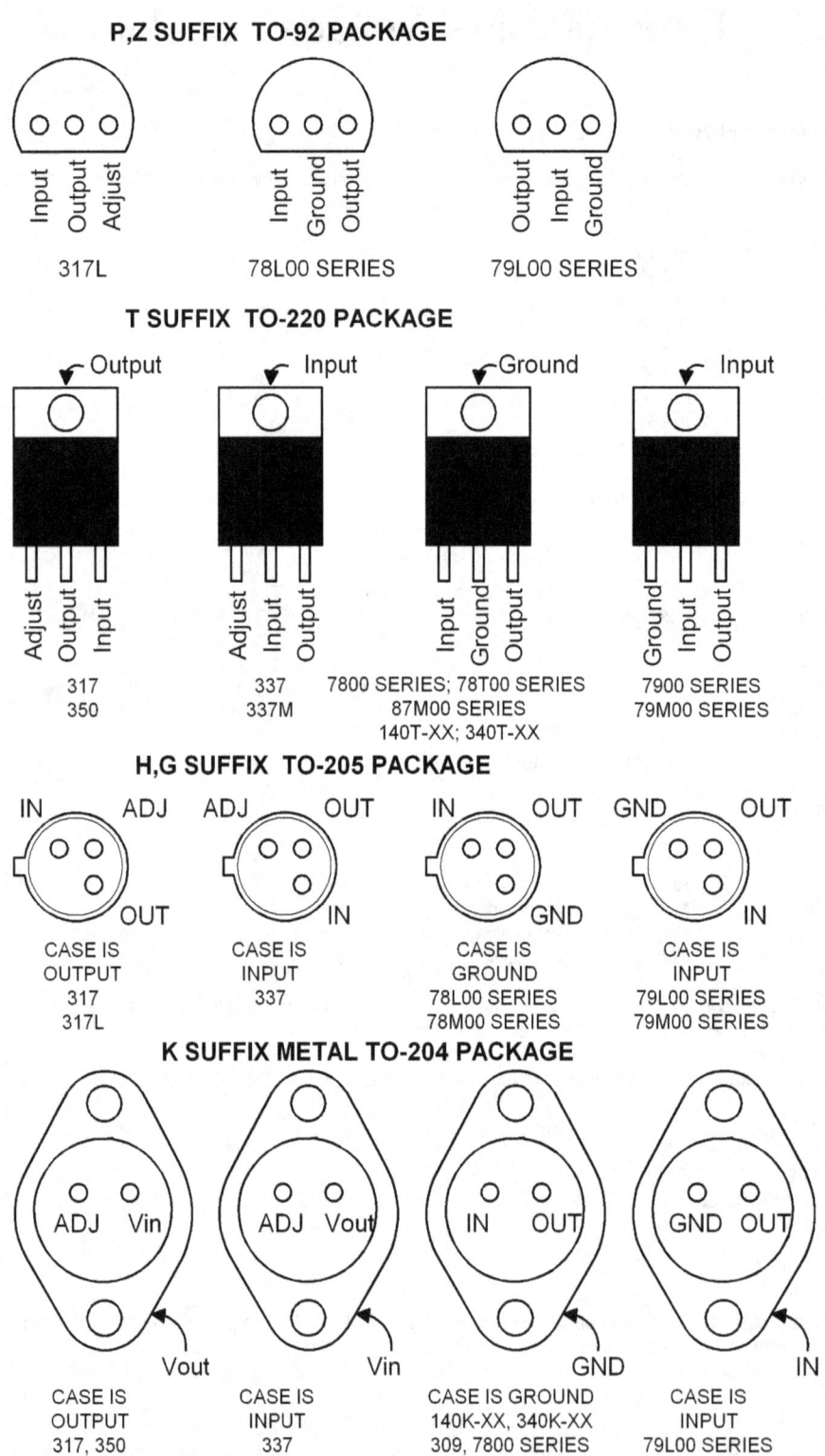

P,Z SUFFIX TO-92 PACKAGE

Input Output Adjust

317L

Input Ground Output

78L00 SERIES

Output Input Ground

79L00 SERIES

T SUFFIX TO-220 PACKAGE

Output

Adjust Output Input

317
350

Input

Adjust Input Output

337
337M

Ground

Input Ground Output

7800 SERIES; 78T00 SERIES
87M00 SERIES
140T-XX; 340T-XX

Input

Ground Input Output

7900 SERIES
79M00 SERIES

H,G SUFFIX TO-205 PACKAGE

IN   ADJ
          OUT
CASE IS
OUTPUT
317
317L

ADJ   OUT
          IN
CASE IS
INPUT
337

IN   OUT
          GND
CASE IS
GROUND
78L00 SERIES
78M00 SERIES

GND   OUT
          IN
CASE IS
INPUT
79L00 SERIES
79M00 SERIES

K SUFFIX METAL TO-204 PACKAGE

ADJ Vin
Vout
CASE IS
OUTPUT
317, 350

ADJ Vout
Vin
CASE IS
INPUT
337

IN OUT
GND
CASE IS GROUND
140K-XX, 340K-XX
309, 7800 SERIES
78T00 SERIES

GND OUT
IN
CASE IS
INPUT
79L00 SERIES

PRINTED CIRCUIT BOARD LAYOUTS

All printed circuit board layouts in this collection are once again printed in the following pages. You can either cut out or photocopy these pages to make a separate file for quick reference.

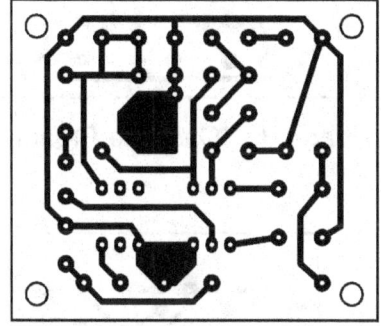

page 13   6.5 W IC Amplifier

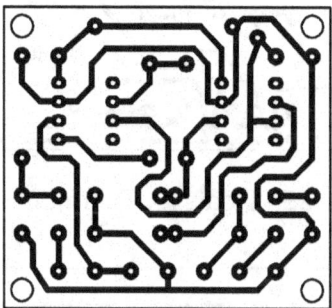

page 19   Guitar Sound Effect

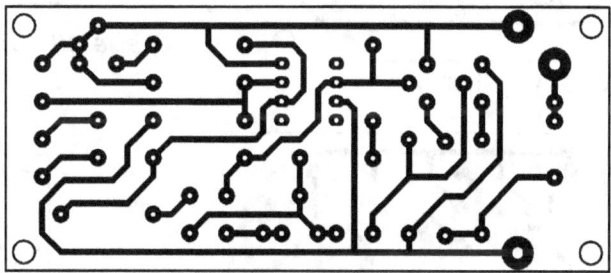

page 15   Mini Amplifier

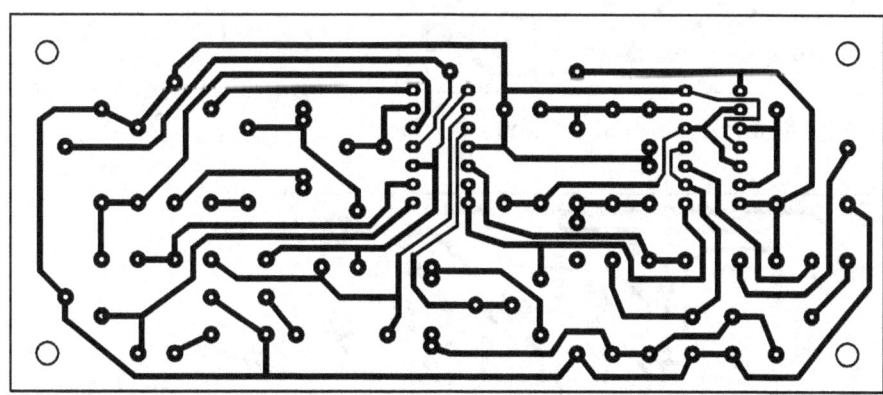

page 21   Audio Squelch

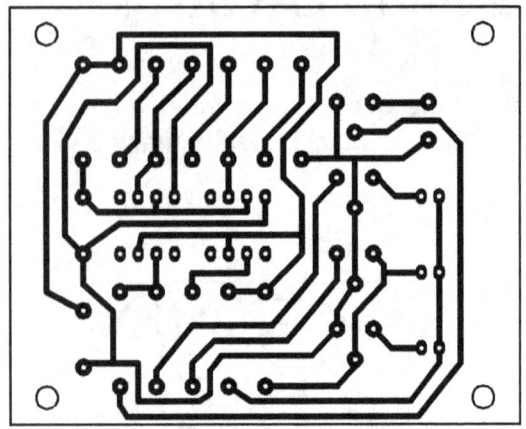

page 25 Heatsink Thermometer

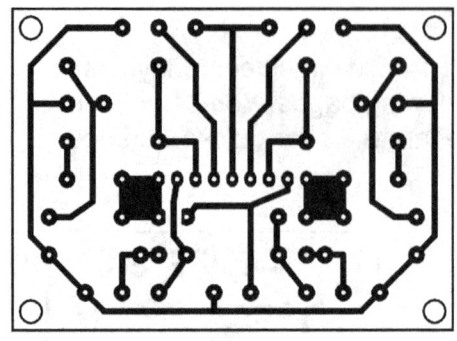

page 33 Cassette Preamp

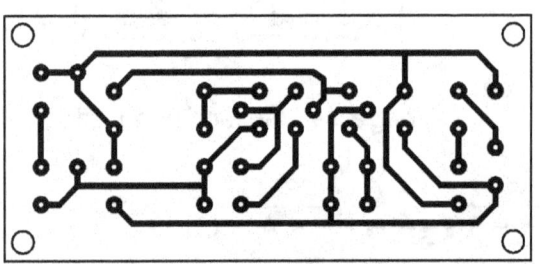

page 29 Video Amplifier

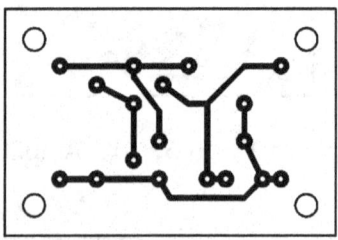

page 40 Variable Zener
Diode

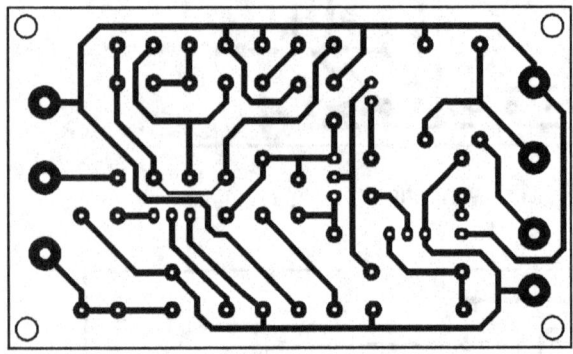

page 31 Automatic Volume Control

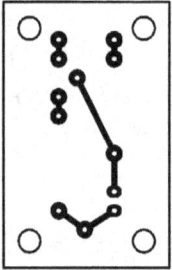

page 41 LED Reference
Diode

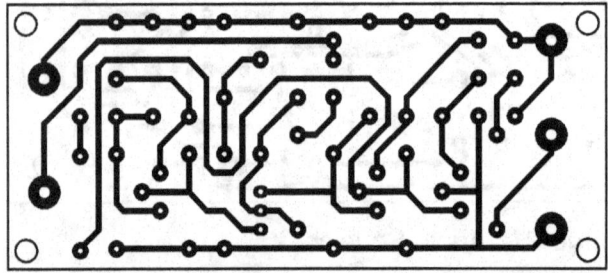

page 37
FM Transmitter

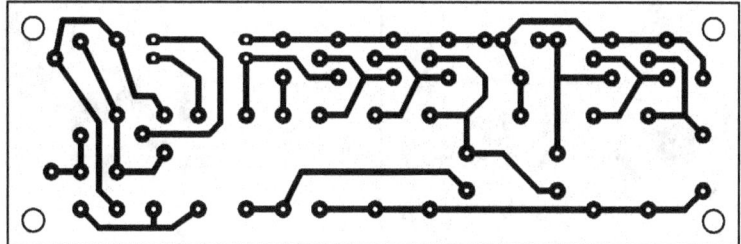

page 39   LED Opto-Coupler

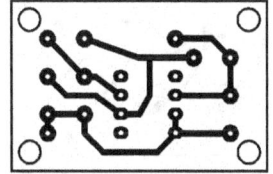

page 44   Infrared Detector

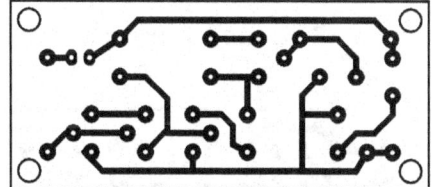

page 45   Dummy Car Alarm

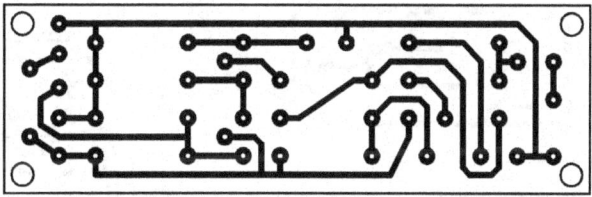

page 49   Housephone Amplifier

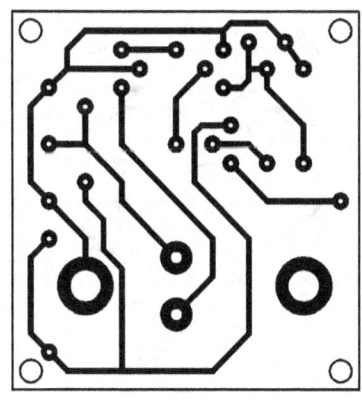

page 51   Touch Doorbell

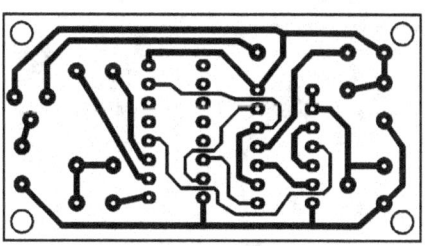

page 54   Universal Beeper

Appendices

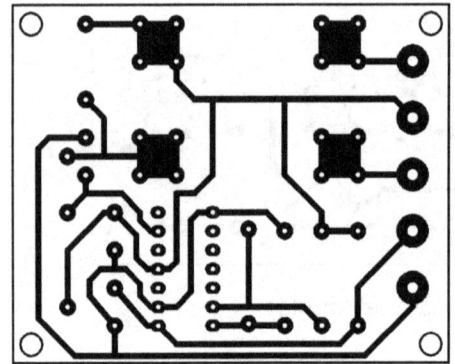

page 56 Double Alarm Circuit

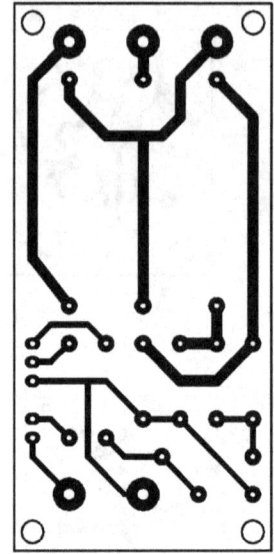

page 61
Car Lamp Monitor

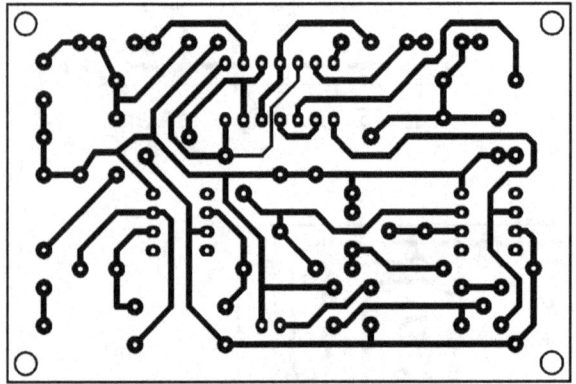

page 59 Telephone Ringer

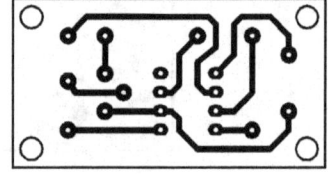

page 66
Multisound Siren

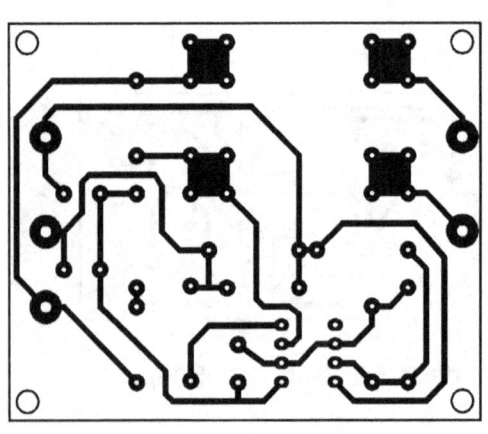

page 63 Car Radio Alarm

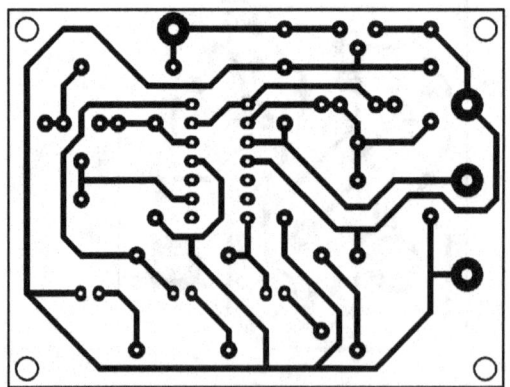

page 65 High Heat Monitor

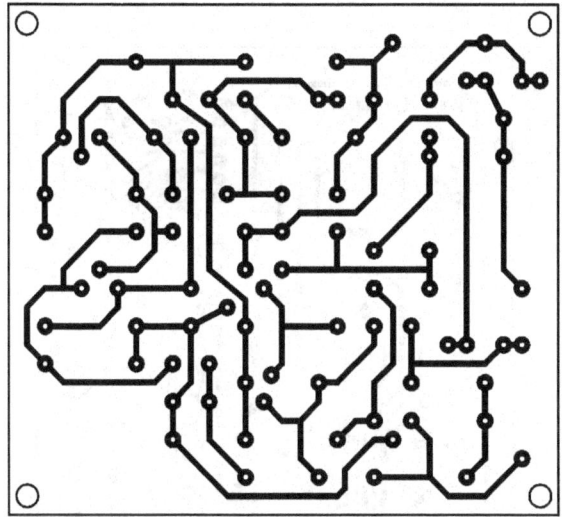

page 71 Mini AM Radio

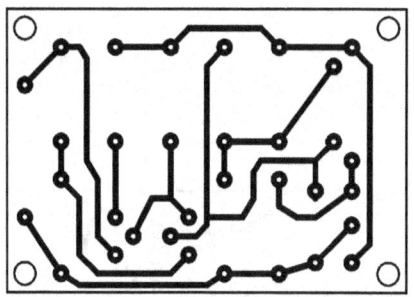

page 76 Video Amplifier

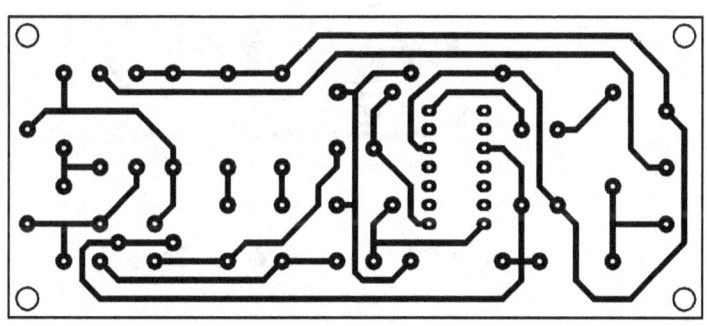

page 75 Tunable Active Antenna

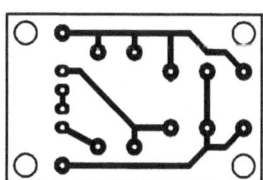

page 82
Model Brake Light

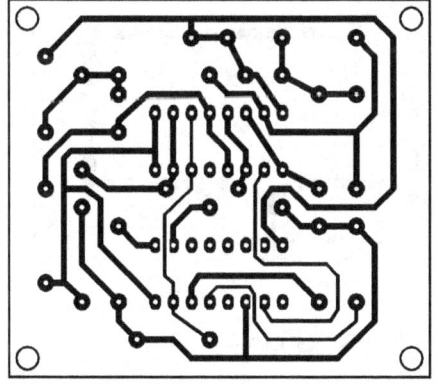

page 79 Repeater Access Encoder

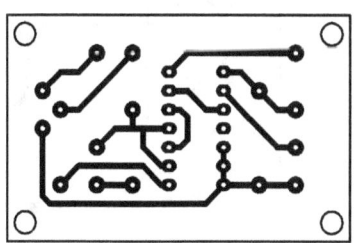

page 86 Quiz Referee

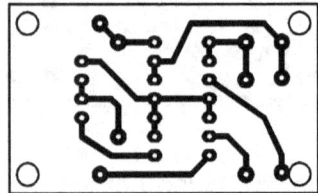

page 93 LED Blinker

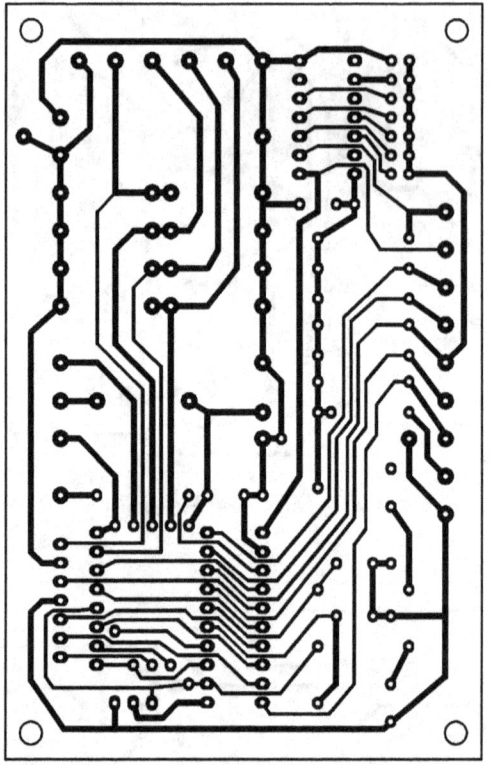

page 96 Stepper Motor Control

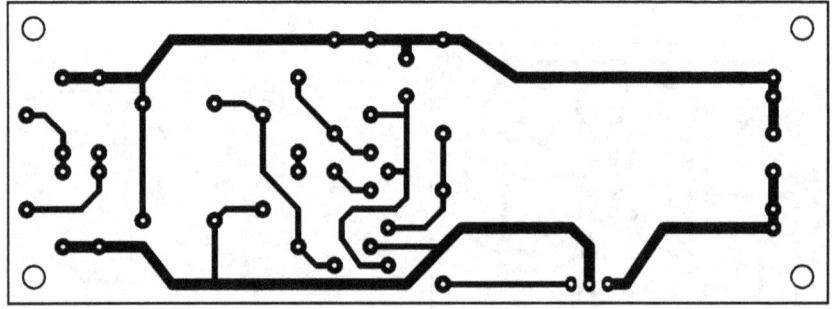

page 101 Stable Power Supply

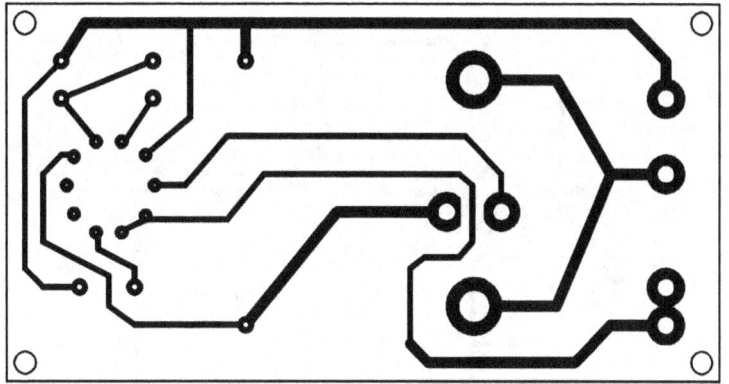

page 103 Constant Current (#1)

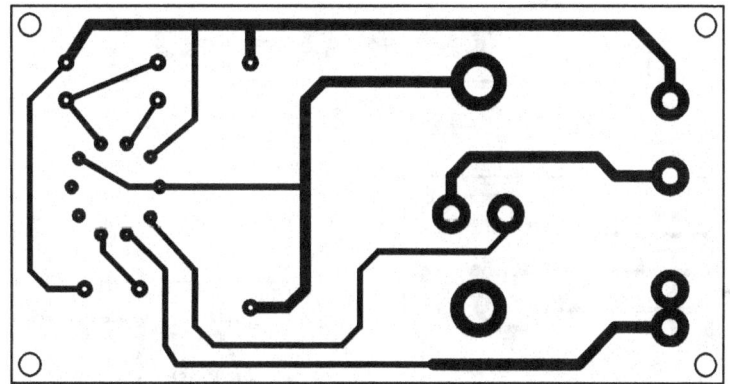

page 104 Constant Current (#2)

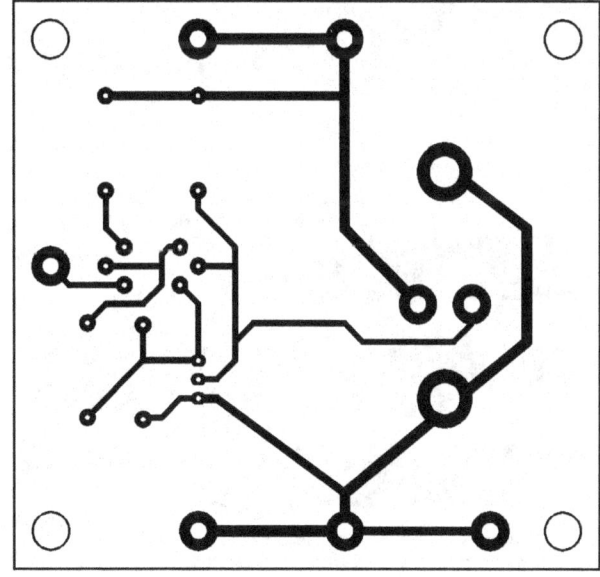

page 106 2N3055 Darlingtons (#1)

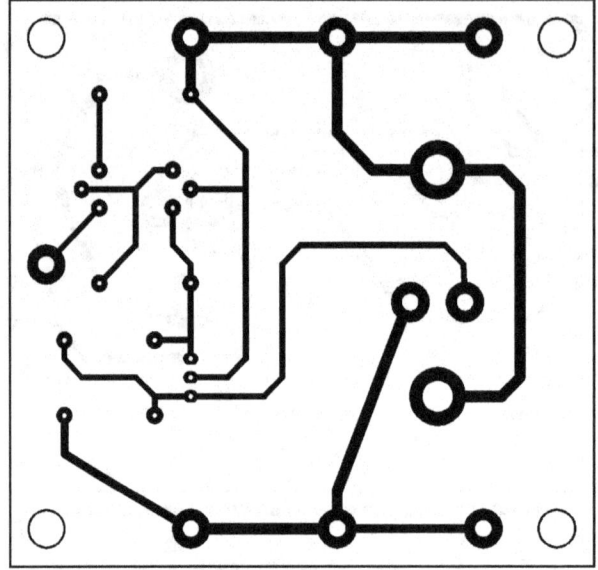

page 107 2N3005 Darlingtons (#2)

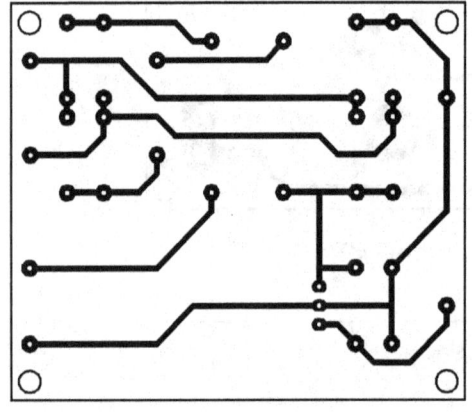

page 110 Standby Supply

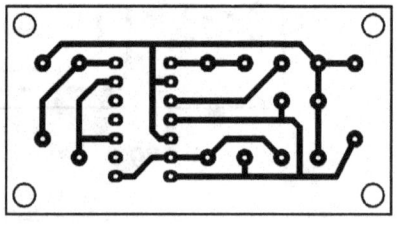

page 115 Polarity Inverter

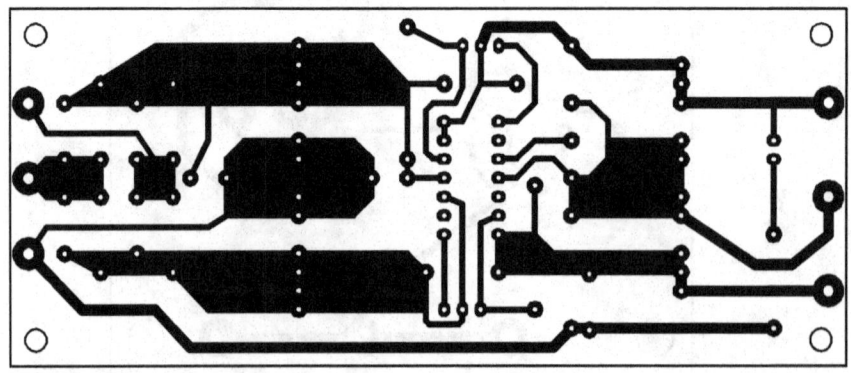

page 116 Supply for Opamps

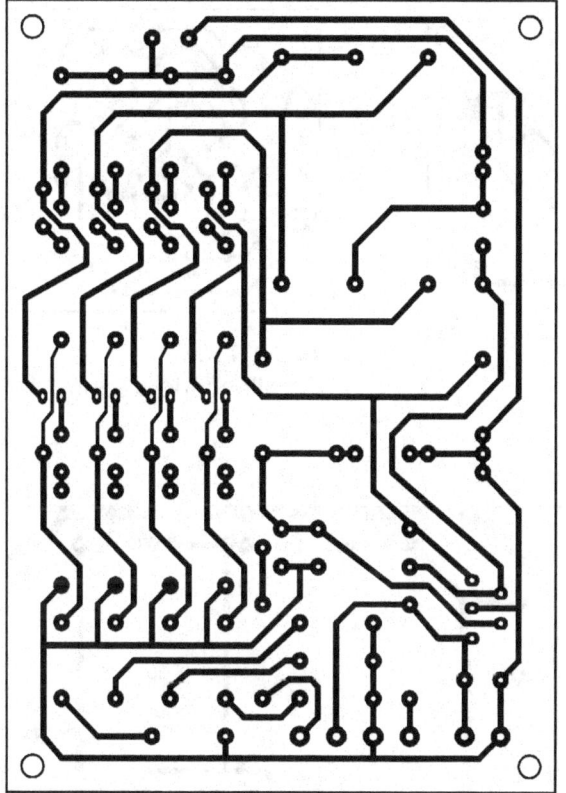

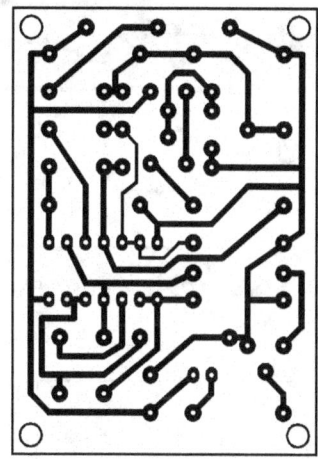

page 131
Signal Injector

page 112   Current Monitored Supply

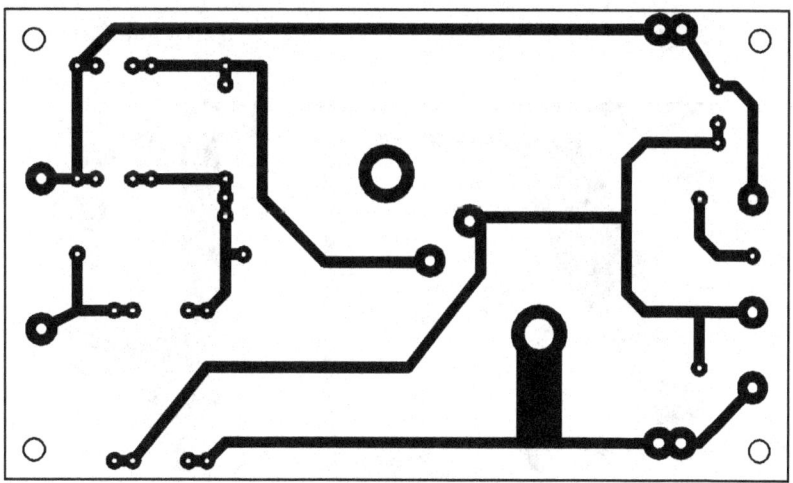

page 120   5A Delayed Power On

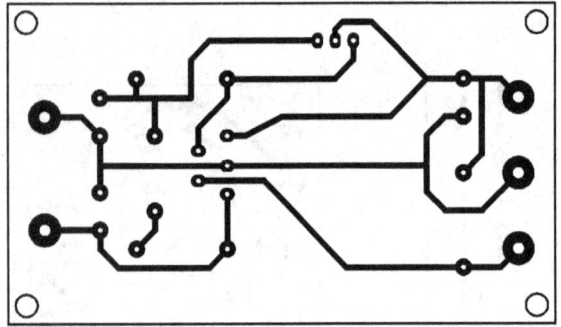

page 122 Robust 5V Supply

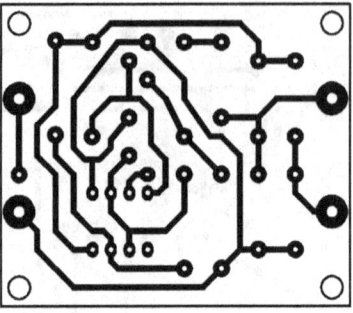

page 135
Oscilloscope Calibrator

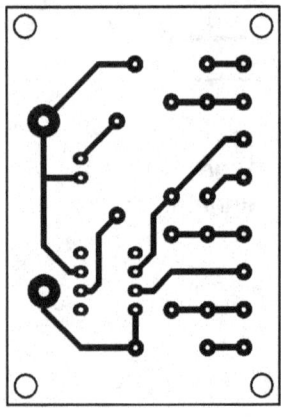

page 139
Transistor Tester

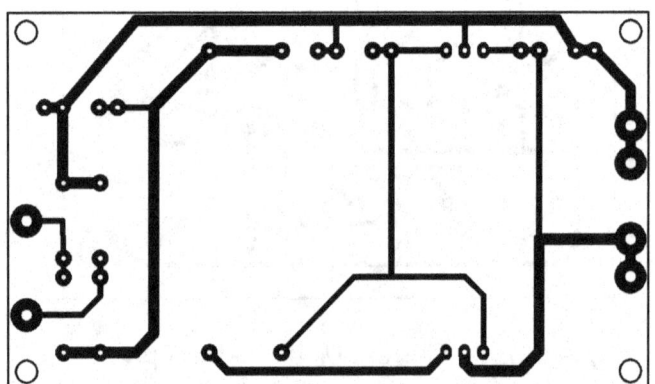

page 124 Amplified Regulator (#1)

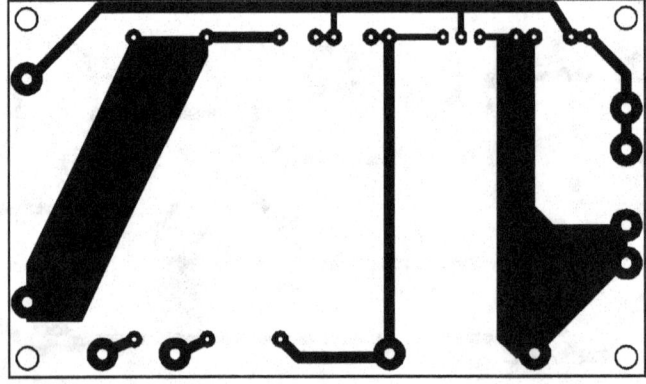

page 125 Amplifier Regulator (#2)

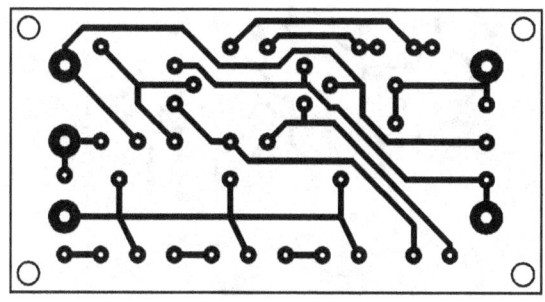

page 129   Linear Ohmmeter

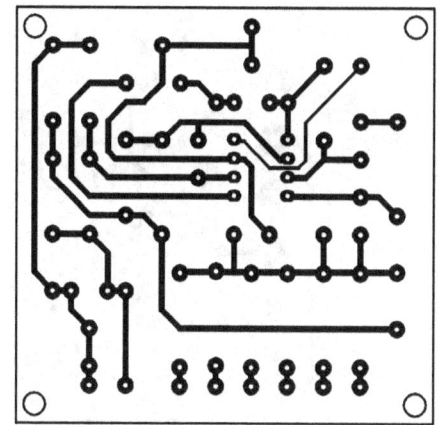

page 133   Wideband Millivoltmeter

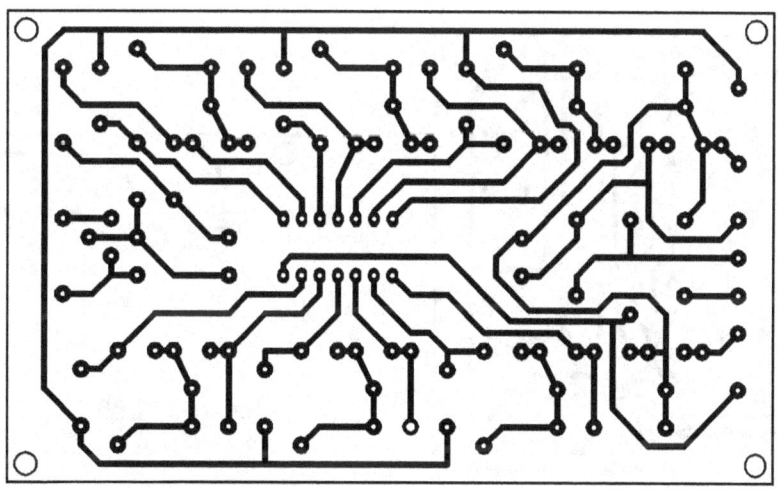

page 137   L & C Meter

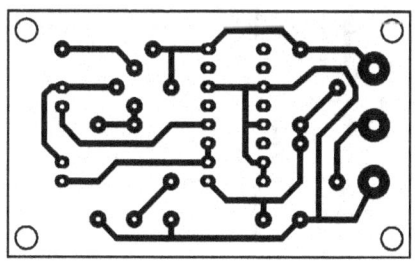

page 140   Adaptive Logic Probe

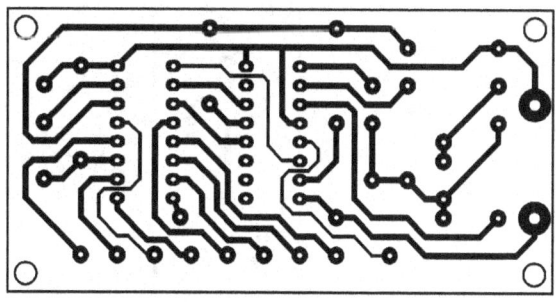

page 142   Oscilloscope Multiplexer

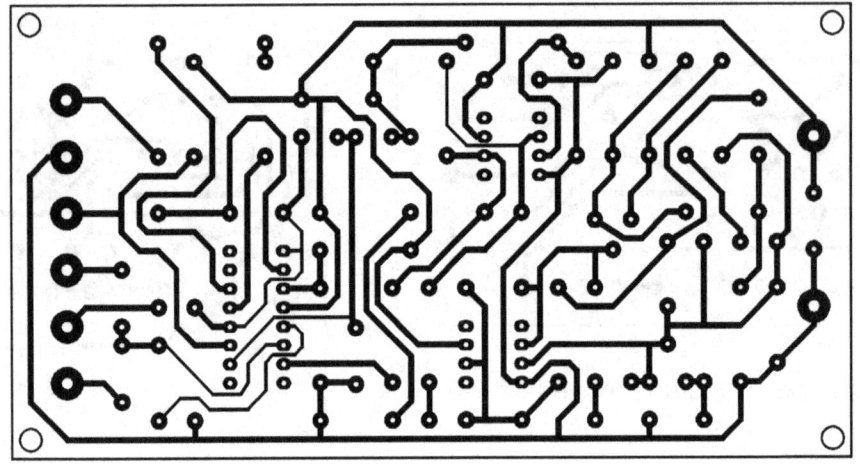

page 144   Sweep Generator

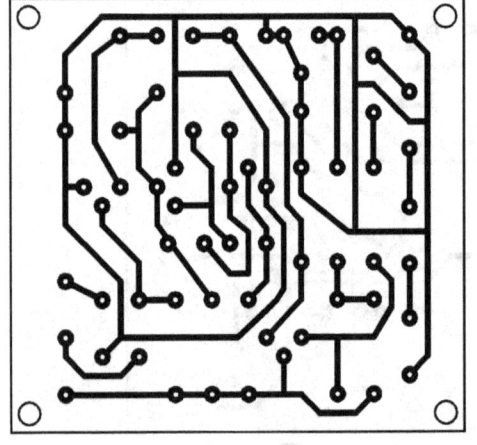

page 152   Audio Scope

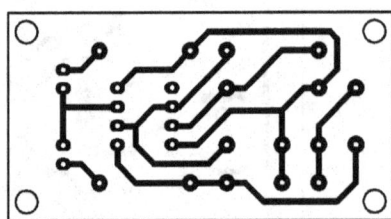

page 150   555 IC Tester

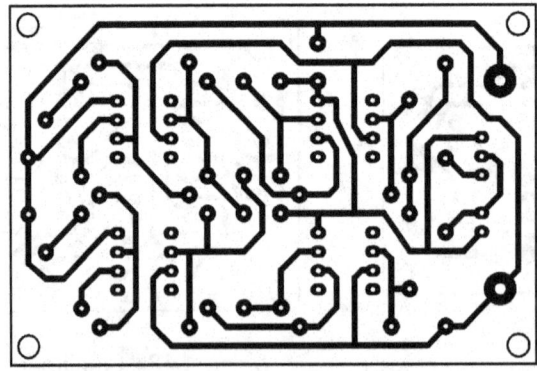

page 148   Voltage Monitor

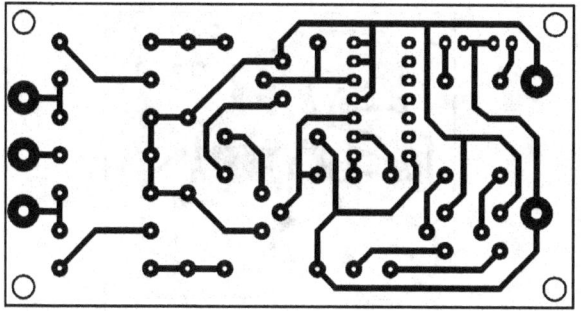

page 154 3-Phase Tester

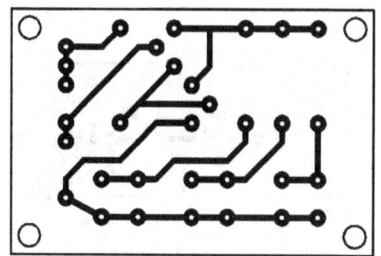

page 156 Zener Diode Tester

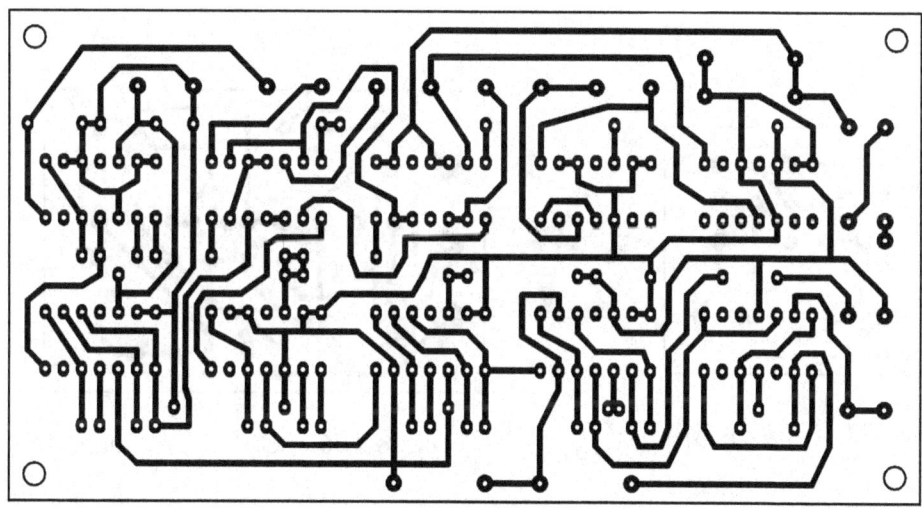

page 159 Digital Clock

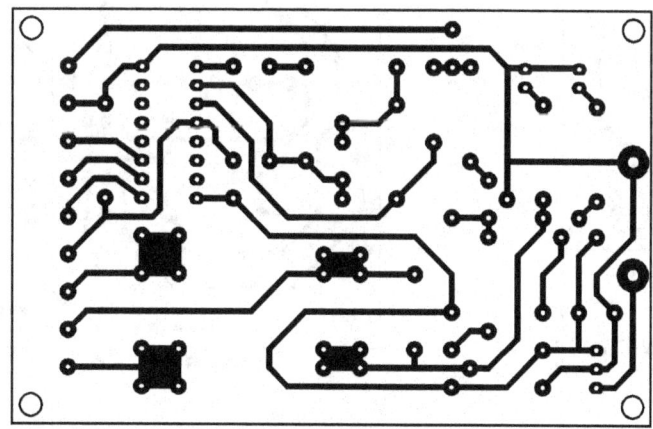

page 164 EPROM IC Eraser

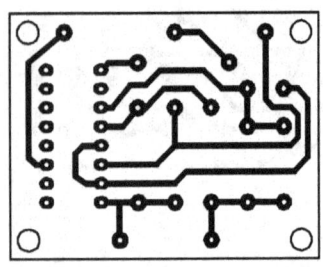

page 179   NE5517 VCO

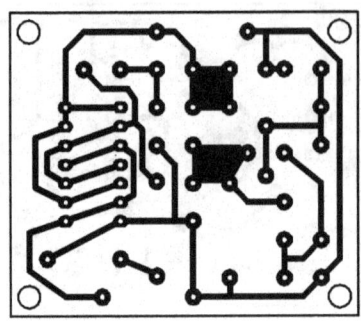

page 180   100 MHz Oscillator

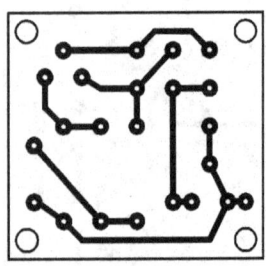

page 183   Morse Keyer

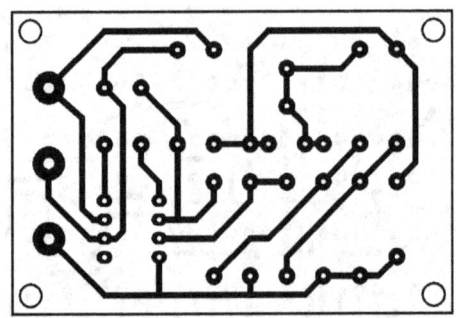

page 186   °C/Voltage Converter

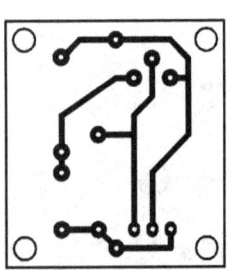

page 189   Power Zener(#1)

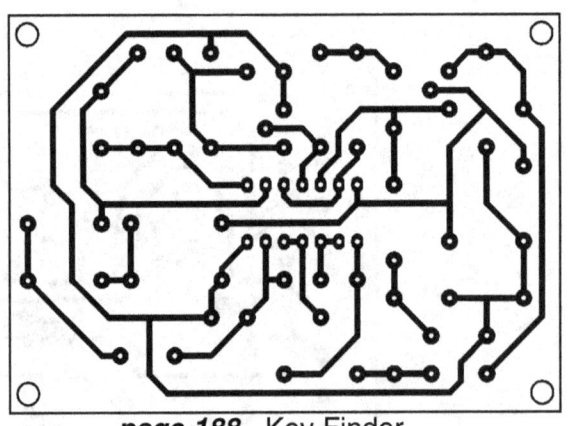

page 188   Key Finder

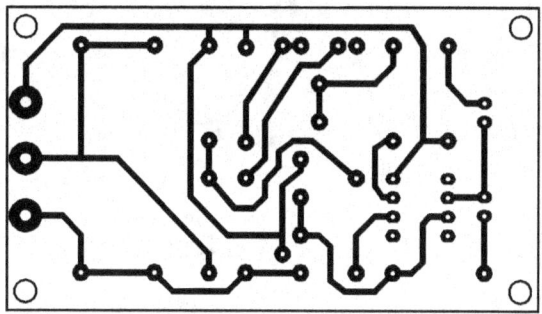

page 191   Charge Monitor

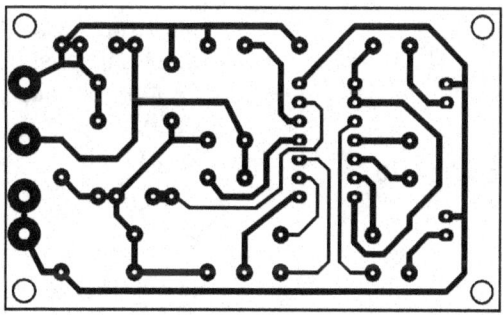

page 193   Temperature Monitor

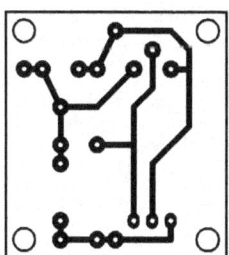

page 190   Power Zener(#2)

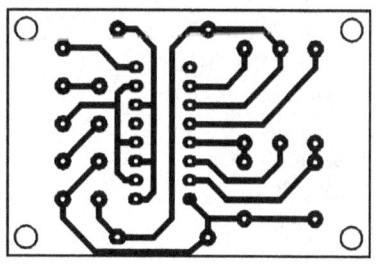

page 194   MOS Signal Injector

Index

Index

Index

www.ingramcontent.com/pod-product-compliance
Lightning Source LLC
Chambersburg PA
CBHW081113170526
45165CB00008B/2436